Springer Series in Experimental Entomology

Thomas A. Miller, Editor

Springer Series In Experimental Entomology
Editor: T.A. Miller

Insect Neurophysiological Techniques
By T.A. Miller
Neurohormonal Techniques in Insects
Edited by T.A. Miller
Sampling Methods in Soybean Entomology
By M. Kogan and D. Herzog
Neuroanatomical Techniques
Edited by N.J. Strausfeld and T.A. Miller
Cuticle Techniques in Arthropods
Edited by T.A. Miller
Measurement of Ion Transport and Metabolic Rate in Insects
Edited by T.J. Bradley and T.A. Miller
Techniques in Pheromone Research
Edited by H.E. Hummel and T.A. Miller
Functional Neuroanatomy
Edited by N.J. Strausfeld

Measurement of Ion Transport and Metabolic Rate in Insects

Edited by
Timothy J. Bradley
Thomas A. Miller

With Contributions by

J.H. Anstee • K. Bowler • T.J. Bradley
D. Brandys • J.W. Hanrahan • A.A. Heusner
J. Machin • S.H.P. Maddrell • L.J. Mandel
J. Meredith • T.A. Miller • M.J. O'Donnell
J.E. Phillips • K. Sláma • M.L. Tracy

With 59 Figures

Springer-Verlag
New York Berlin Heidelberg Tokyo

Timothy J. Bradley
Department of Developmental
and Cell Biology
School of Biological Sciences
University of California
Irvine, California 92717
USA

Thomas A. Miller
Department of Entomology
University of California
Riverside, California 92521
USA

Library of Congress Cataloging in Publication Data
Main entry under title:
Measurement of ion transport and metabolic rate in
insects.
Bibliography: p.
Includes index.
1. Insects—Physiology. 2. Biological transport.
3. Epithelium. 4. Entomology—Methodology. I. Miller,
Thomas A. II. Bradley, T. J. (Timothy J.)
QL495.M4 1983 595.7'01858 83-6683

Softcover reprint of the hardcover 1st edition 1984

Typeset by MS Associates, Champaign, Illinois.

9 8 7 6 5 4 3 2 1

ISBN-13:978-1-4613-8241-6 e-ISBN-13:978-1-4613-8239-3
DOI: 10.1007/978-1-4613-8239-3

Series Preface

Insects as a group occupy a middle ground in the biosphere between bacteria and viruses at one extreme, amphibians and mammals at the other. The size and general nature of insects present special problems to the student of entomology. For example, many commercially available instruments are geared to measure in grams, while the forces commonly encountered in studying insects are in the milligram range. Therefore, techniques developed in the study of insects or in those fields concerned with the control of insect pests are often unique.

Methods for measuring things are common to all sciences. Advances sometimes depend more on how something was done than on what was measured; indeed a given field often progresses from one technique to another as new methods are discovered, developed, and modified. Just as often, some of these techniques find their way into the classroom when the problems involved have been sufficiently ironed out to permit students to master the manipulations in a few laboratory periods.

Many specialized techniques are confined to one specific research laboratory. Although methods may be considered commonplace where they are used, in another context even the simplest procedures may save considerable time. It is the purpose of this series (1) to report new developments in methodology, (2) to reveal sources of groups who have dealt with and solved particular entomological problems, and (3) to describe experiments which may be applicable for use in biology laboratory courses.

THOMAS A. MILLER
Series Editor

Contents

1. Introduction
T.J. Bradley and T.A. Miller 1

2. *In Vitro* Techniques for Studies of Malpighian Tubules
M.J. O'Donnell and S.H.P. Maddrell
With 3 Figures 5

3. Methods for the Study of Transport and Control in Insect Hindgut
J.W. Hanrahan, J. Meredith, J.E. Phillips, and D. Brandys
With 8 Figures 19

4. The Study of Atmospheric Water Absorption
J. Machin
With 11 Figures 69

5. Microrespirometry in Small Tissues and Organs
K. Sláma
With 6 Figures 101

6. Spectrophotometry and Fluorometry in Ion Transport Epithelia
Lazaro J. Mandel
With 14 Figures 131

7. Coulometric Measurement of Oxygen Consumption in Insects
A.A. HEUSNER and M.L. TRACY
With 11 Figures 163

8. Techniques for Studying Na^+, K^+ -ATPase
J.H. ANSTEE and K. BOWLER
With 6 Figures 187

Index 221

Contributors

J.H. Anstee
Department of Zoology, University of Durham, Durham DH1 3LE, England

K. Bowler
Department of Zoology, University of Durham, Durham DH1 3LE, England

T.J. Bradley
Department of Developmental and Cell Biology, School of Biological Sciences, University of California, Irvine, California 92717, USA

D. Brandys
Department of Zoology, University of British Columbia, Vancouver, British Columbia V6T 1W5, Canada

J.W. Hanrahan
Department of Zoology, University of British Columbia, Vancouver, British Columbia V6T 1W5, Canada

A.A. Heusner
Department of Physiological Sciences, School of Veterinary Medicine, University of California, Davis, California 95616, USA

J. Machin
Department of Zoology, University of Toronto, Toronto, Ontario M55 1A1, Canada

S.H.P. MADDRELL
Department of Zoology, University of Cambridge, Cambridge CB2 3EJ, England

LAZARO J. MANDEL
Department of Physiology, Duke University Medical Center, Durham, North Carolina 27710, USA

J. MEREDITH
Department of Zoology, University of British Columbia, Vancouver, British Columbia V6T 1W5, Canada

T.A. MILLER
Department of Entomology, University of California, Riverside, California 92521, USA

M.J. O'DONNELL
Department of Zoology, University of Cambridge, Cambridge CB2 3EJ, England

J.E. PHILLIPS
Department of Zoology, University of British Columbia, Vancouver, British Columbia V6T 1W5, Canada

K. SLÁMA
Department of Insect Physiology, Institute of Entomology CSAV, Praha 6, Czechoslovakia

M.L. TRACY
Department of Physiological Sciences, School of Veterinary Medicine, University of California, Davis, California 95616, USA

Chapter 1
Introduction

T. J. Bradley and T. A. Miller

Arthropods, particularly the insects, offer distinct advantages as sources of experimental material for the study of ion and fluid transport. Insect epithelia are histologically simple. They are one cell layer thick and generally not associated with underlying cellular connective tissues. These characteristics have contributed to the substantial successes to date in isolating intact and viable epithelial organs for study *in vitro.* The chapters in this volume by O'Donnell and Maddrell (Chapter 2) and by Hanrahan et al. (Chapter 3) demonstrate the elegant techniques that have been developed to exploit these advantages. Insect physiologists have pioneered many miniaturizations of analytical techniques in response to the challenges of working with small animals. Machin (Chapter 4) describes in detail a number of the techniques that he has developed to analyze the uptake of water from subsaturated air by small arthropods.

In other areas of ion transport physiology, our information on insect tissues has lagged behind that of workers using other organisms. One clear example of this is in the area of the biochemistry of ion-transporting enzyme complexes. The difficulty of obtaining large quantities of isolated, homogeneous insect tissue has complicated the isolation and characterization of the metabolic units responsible for ion translocation. One notable exception to this has been the work on Na^+, K^+-ATPase in the insects, a topic reviewed by Anstee and Bowler (Chapter 8).

The difficulty of investigating insect transport systems through biochemical analysis has led some physiologists to examine the links between transport and cell metabolism. One of the attractions of ion and water transport as physiological phenomena for study is the capacity to place these events in a physical context; that is, ion gradients and osmotic gradients can be equated with units of potential energy that have been produced and stored by metabolic events in the membranes. These concepts have long been used by membrane physiologists to determine whether transport events are active or passive. In the strictest thermodynamic sense, transport processes that lead to a net increase in the potential energy stored as electrical, osmotic, and pressure gradients can be considered to be active even if one is unaware of the biochemical mechanisms by which these have arisen.

Recently, it has become apparent that these concepts of the storage of potential energy across membranes are useful not only as theoretical "black box" approaches to transport processes and tissues, but also as accurate descriptions of important membrane functions. In particular, the Mitchell hypothesis (Mitchell 1966), which proposes a direct link between the H^+ gradient generated across bacterial, mitochondrial, and chloroplast membranes, and the ability of these membranes to produce ATP, has demonstrated the role of membranes as organelles of energy storage. Similar roles could be argued for the plasma membranes of nerve and muscle which store energy in the Na^+ and K^+ gradients for use in propagating action potentials, or for the membranes of intestinal and kidney cells in the vertebrates and Malpighian tubule cells in the insects which use energy stored in the Na^+ gradient to transport amino acids against their concentration gradients.

The Editors feel that investigations of linkages between transport and cellular metabolism will continue to be a fruitful and important area for study in the field of insect physiology. A number of investigators have pointed out the possibility that the mechanisms by which energy is provided to transport processes in the insects may differ from those known in the vertebrates (Bradley 1983; Harvey et al. 1981; Keynes 1973). For these reasons the Editors have seen fit to combine, in one volume, descriptions of current techniques for the measurement of ion transport, water movements, and metabolic rate.

Conventionally, energy metabolism is measured in the form of oxygen uptake or carbon dioxide evolution. Miniaturizations of standard techniques are required for the study of some of the smaller insects or isolated insect organs. Sláma (Chapter 5) describes a simple and inexpensive volumetric respirometer suitable for either teaching or research applications, as well as another more sensitive unit appropriate for detailed, prolonged studies on respiration. Heusner and Tracy (Chapter 7) present a detailed description of an apparatus for the coulometric measurement of oxygen consumption. The quantitative replacement of oxygen in the respirator allows this instrument to combine high resolution with the capacity for long-term use. For some applications, techniques are required that provide simultaneous measurements of ion transport and metabolic rates with very high time resolution. Mandel (Chapter 6) describes the use of spectrophotometry and fluorometry to monitor the redox state of respiratory enzymes in isolated epithelial tissues.

It is our hope that students of insect physiology attracted to this volume out of an interest in one of the techniques will recognize the possible significance of all the techniques in their work. If physiologists interested in metabolism and energy flow within insects are led by the juxtaposition of these chapters to speculate about the cost of ionic and osmotic regulation in their organisms, and if transport physiologists are led to analyze metabolic phenomena in their preparations, our efforts will have been richly rewarded.

References

Bradley TJ (1983) Mitochondrial placement and function in insect ion-transporting cells. Am Zool (in press)

Harvey WR, Cioffi M, Wolfersberger MO (1981) Portasomes as coupling factors in active transport and oxidative phosphorylation. Am Zool **21**:775–791

Keynes RD (1973) Comparative aspects of transport through epithelia. In: Ussing HH, Thorn NA (eds) *Transport mechanisms in epithelia*. Academic Press, New York

Mitchell P (1966) Chemiosmotic coupling in oxidative and photosynthetic phosphorylation. Biol Rev **41**:445–502

Chapter 2

In Vitro Techniques for Studies of Malpighian Tubules

M. J. O'Donnell and S. H. P. Maddrell

I. Experimental Advantages of Malpighian Tubules for *In Vitro* Studies of Epithelial Function

Insect Malpighian tubules are extrordinarily useful for studies of virtually all facets of epithelial ion and water transport. Their greatest experimental advantage results from their geometry: tubules consist of a single layer of squamous epithelial cells that form a blind-ended cylinder. Secretion products, which form the luminal contents of the cylinder, can thus easily be kept separate from the bathing fluids *in vitro*. If a drop of bathing fluid is placed in 5-6 mm paraffin oil on wax, and the open end of the tubule is pulled out of the drop, the secretion issuing from the open end will form a second droplet that will adhere to a steel or glass pin by surface tension (Ramsay, 1954). The volume of the spherical droplet can be calculated from the formula $4/3\pi r^3$, where r, the radius of the drop, is measured by an eyepiece graticule fitted to a dissecting microscope. Such straightforward methods circumvent much of the experimental interference and the need for elaborate experimental chambers entailed when studying fluid or solute movements across a flat sheet of epithelial cells.

There are also a great many economic advantages in using Malpighian tubules in studies of epithelial transport. Many insects can be cultured cheaply and in large numbers, and there are typically 4–6 tubules per individual, although some species have as many as 200 tubules per individual. More importantly, the small size of tubules (50–100 μm in diameter and 3 cm in length in *Rhodnius*) means that very small quantities of bathing fluids are required for experiments, and the use of expensive pharmacological reagents or radioisotopes is therefore many times less than in comparable studies of other epithelia.

The ease of dissection means that time, and hence money, is saved when tubules are prepared for research or teaching. The necessary dissection technique can be acquired with a few hours of practice, although for teaching purposes tubules are usually dissected by experienced personnel and set up in paraffin oil for the students. We have found that two laboratory demonstrators can easily supply up to 12 students with as many as 12 tubules each during a 5-h laboratory session.

The small size of tubules and their large surface area relative to their volume allows tubules to respond rapidly to experimental changes in the composition of the bathing fluids. The size of the tubules does not mean, however, that secretion rates or the composition of the secretion is difficult to measure, because some tubules secrete *in vivo* at rates up to 0.48 ml min^{-1} $tissue^{-1}$, the highest yet recorded for any tissue. *In vitro,* maximal secretion rates (50 nl min^{-1} $tubule^{-1}$) in *Rhodnius* are obtained by stimulating the tubules either with diuretic hormone extracted from the mesothoracic ganglionic mass (Maddrell 1964), or with the cheap and readily available diuretic hormone mimic 5-hydroxytryptamine (5-HT) at a concentration of 10^{-6} *M* (Maddrell et al., 1969). Slower secretion rates can be achieved through stimulation by suboptimal doses of 5-hydroxytryptamine (10^{-8} - 10^{-7} *M*), or through addition of appropriate quantities (10^{-5} - 10^{-4} *M*) of a competitive inhibitor, tryptamine (Maddrell et al., 1971).

An additional advantage of using Malpighian tubules for epithelial studies is that much is already known about the structure and function of Malpighian tubules (see references cited by Maddrell 1980b). This knowledge is a useful foundation for further examinations of pharmacological and histological properties of tubules, the mechanism by which fluid secretion is driven by ion pumping, and the role of the tubules in excretion and detoxification. Studies in the latter areas are of course relevant to pest management programs using insecticides.

In this chapter experimental techniques are described that have proven useful in the study of Malpighian tubules. A brief description of dissection techniques for various species precedes a more detailed discussion of methods for perfusing the apical (luminal) and basic (serosal) surfaces of tubules. Methods for measuring the electrical properties of tubules are then presented.

II. Dissection of Malpighian Tubules

1. Tubules from *Rhodnius*

The dissection technique for *Rhodnius* and the method for measuring secretion *in vitro* have been discussed in detail in a previous volume in this series (Maddrell 1980a). In brief, the method is as follows. A fifth instar *Rhodnius,* 2-3 weeks post-molt, is placed dorsal surface uppermost in a wax dissecting dish and sacrificed by crushing the head with fine-pointed forceps. The carcass is secured with bent pins placed so that they hold down the outstretched legs on either side of the abdomen. The dorsal surface of the abdomen is first removed with a fine scalpel and forceps. The crop and midgut are then removed: the crop is grasped with forceps and pulled anteriorly, while the midgut is pulled with a second pair of forceps at the point where the Malpighian tubules join the gut. Glass rods, pulled over a low flame so that their tips are 200-300 μm in diameter, are used to unravel the coiled mass of Malpighian tubules. Two glass rods are placed in a loop of tubule and pulled in opposite directions so as to expand the loop and break the adherent tracheae. Once a tubule is unraveled, the upper, fluid-secreting portion can be removed by cutting it just beyond the junction with the lower portion of the tubule. The junction is readily discernible by the different appearances of upper and lower tubule: the wall of the upper tubule is opaque and the lumen contains a colorless fluid, whereas the wall of the lower tubule is translucent and the lumen, for much of its length, contains numerous white crystals of uric acid.

2. Tubules from Other Insects

The tubules of ***Rhodnius*** are comparatively easy to dissect because they are fairly elastic and their adherent tracheae are quite easily removed. In other species the tubules can be more brittle and the tracheae tougher and more numerous; the dissection, therefore, is more difficult. Nevertheless, the tubules of many species have been successfully removed and set up *in vitro*, and experience suggests that many as yet unstudied species will be amenable to the dissection techniques described below.

The original technique for dissecting tubules and studying their secretion *in vitro* was developed by Ramsay (1954) for the stick insect *Carausius morosus*. The specimen is pinned ventral surface uppermost under dissecting fluid, and the abdomen is then slit lengthwise and pinned open. The tubules are removed much as for ***Rhodnius***.

To remove tubules of dipterans such as the tsetse fly *Glossina* or the blowfly *Calliphora*, the insect is pinned through the thorax, ventral surface uppermost. One of the anterior abdominal sclerites is grasped with forceps, pulled one abdomen length posteriorly, and pinned to the dish. The abdominal contents spill out through the ruptured cuticle, and the gut can be pulled to one side to facilitate dissection of the tubules, which are removed much as for ***Rhodnius***. However, since the tubules are more brittle, it is advisable to sever as many tracheal connections as possible using fine scissors. For small dipterans, such as mosquito larvae, the tubules are often only 3 mm in length, and therefore only 50-60% of the tubule can be covered by the bathing solution. Such tubules secrete only about 250 pl min^{-1}. To collect sufficient material for analysis, Phillips and Maddrell (1974) suggest that up to five tubules be included in each drop, each tubule delivering fluid to one of a row of glass collecting rods set in an arc close to the bathing drop.

Tubules from larval lepidopterans such as ***Pieris brassicae*** and ***Manduca sexta*** are readily accesible. The larva is pinned through head and tail, the dorsal surface slit lengthwise and pinned open, and the tubules dissected free of tracheal connections as in *Rhodnius.*

Dissection of adult butterflies is somewhat more difficult and has been described in detail by Nicolson (1976). The abdomen of the insect is opened under dissecting medium and the midgut is pulled gently away from the body. This procedure frees most of the tubules with minimum handling. Glass rods are used to separate the distal ends of the tubules from the fat body around the rectum.

A quite different technique has been developed for preparing tubules of the desert locust, *Schistocerca gregaria* (Maddrell and Klunsuwan 1973). This species has about 200 tubules that form a densely tangled mass around the alimentary canal, making isolation of individual tubules quite difficult. The whole alimentary canal is therefore removed from the insect. The abdomen is first severed cleanly with scissors about 5 mm from the posterior tip. The head of the insect is held in one hand and pulled smoothly and directly away from the thorax, held in the other hand. The thin cuticle of the neck tears, and the gut with adherent Malpighian tubules is drawn through the neck. The severed head of the hindgut is ligated with a silk thread to prevent escape of the gut contents. The entire gut is then immersed in 2-3 ml Ringer's solution contained in a wax trough under liquid paraffin. Small staples or bent pins are used to tether the head forward of the Ringer's, thus preventing contamination of the preparation by regurgitated fluid. Any tubules so contaminated do not function. Finally, individual Malpighian tubules are cut close to their point of entry into the alimentary canal, and the cut ends are pulled out of the Ringer's to fine glass rods stuck into the wax adjacent to the Ringer's solution bath. Advantages of this technique are that the tubules are handled minimally, residual oxygen in the uninterrupted tracheae may benefit tubule functioning, and as many as 30 tubules from one insect can be run simultaneously.

III. Perfusion of Malpighian Tubules

1. Perfusion of the Tubule Lumen

Much of our knowledge of tubule function is based on experiments in which the composition and/or flow rate of the luminal fluids is varied. For example, it is often of interest to measure the effects of a particular ion concentration or pharmacological agent on the apical (i.e., luminal) cell membranes, or to measure the flux of a particular solute in an apical to basal direction. A technique for continuous perfusion of the lumen of Malpighian tubules has therefore been an important experimental tool. The technique is also important in cases in which tubules secrete fluid at low rates. In such cases, the transport of materials into the lumen can still be studied by running fluid through the lumen so as to carry out the luminal fluid.

The first step of the procedure is to place a dissected tubule in a drop of Ringer's solution in 4-5 mm paraffin oil. The oil sits on wax lining a 10-cm petri dish that is illuminated from below (Fig. 2.1a). A pair of fine (No. 5) forceps are clamped to a three-axis micromanipulator, moved adjacent to the drop, and

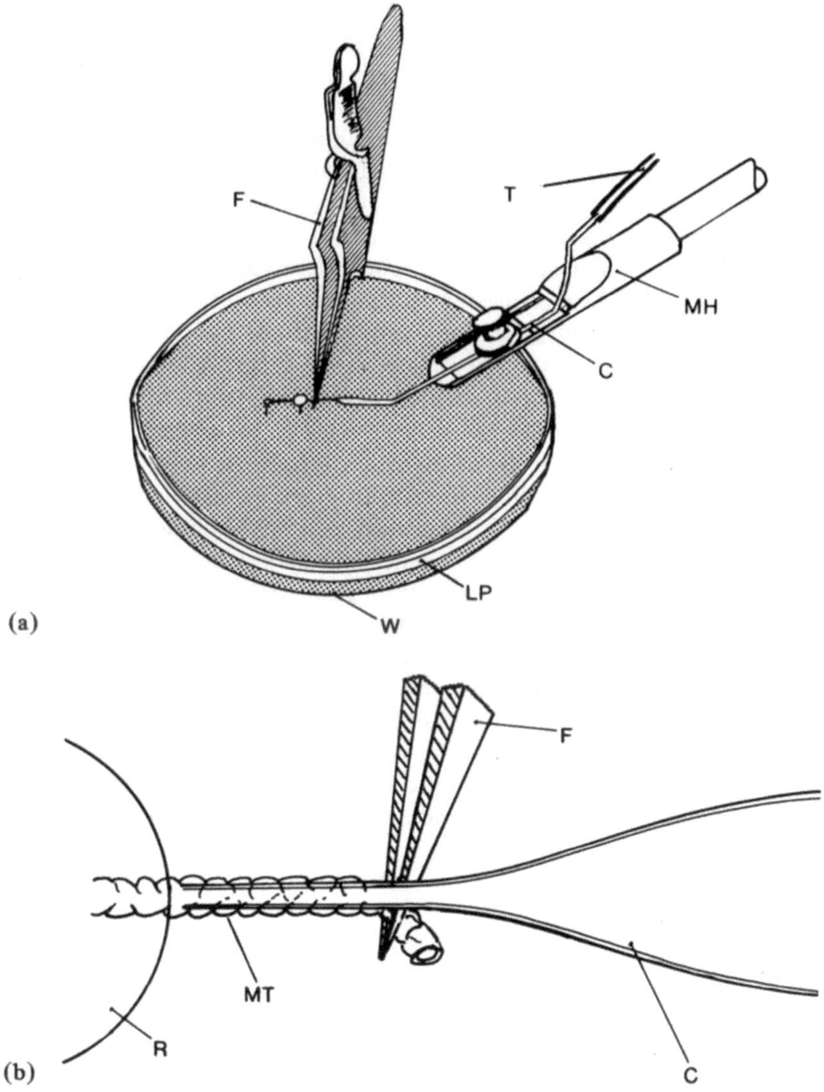

Fig. 2.1. Apparatus used for perfusing the lumen of a Malpighian tubule. **(a)** General view. C, cannula; F, fine-pointed forceps; LP, liquid paraffin; MH, microelectrode holder; T, thick-walled tubing; W, translucent wax. See text for procedure. **(b)** Close-up of the cannula (C) entering the Malpighian tubule (MT) in a drop of Ringer's solution (R).

arranged so that the tips open and close in the horizontal plane. The tubule is wrapped one or two times around the nearest point of the forceps, which are then held closed with a small spring clamp or alligator clip. A cannula is formed by breaking the tip from a glass micropipette to produce an opening about 30 μm in diameter. If filamented glass is used, the cannula can be easily back-filled with physiological saline or experimental solution. The cannula is held in a microelectrode holder mounted in a second micromanipulator. The micromanipulator is used to position the cannula with its long axis in line with the length of tubule, and with the cannula's tip pointing at right angles to the forceps holding the tubule. The cannula is first advanced so that its tip penetrates the tubule wall, and is then further advanced until the taper of the cannula fills the tubule lumen. Practice is required to avoid tearing the tubule away from the forceps or advancing the cannula through the opposite wall of the tubule.

Fluid is perfused through the tubule by a motor-driven micrometer syringe (e.g., Agla Micrometer Syringe outfit, Wellcome Research Laboratories, Beckenham, Kent, U.K.), which is connected to the cannula by thick-walled flexible tubing. The syringe, usually a Hamilton type of 50- or 100-μl capacity, is attached by a rigid holder to the micrometer screw gauge. A variable-speed DC motor and reduction gearhead (Portescap, Reading, U.K.) turns the screw gauge, which advances a spindle against the syringe plunger. The perfusion rate is linearly related to the rotation speed of the screw micrometer gauge; the system is easily calibrated by collecting drops of fluid from the syringe at timed intervals for a series of rotation speeds. Thick-walled polyvinyl chloride tubing (e.g., 1.1 mm i.d., 4.7 mm o.d.) is used to connect the syringe and cannula; in thin-walled tubing some fluid may evaporate through the tubing's wall. The syringe, tubing, and cannula must also be ascertained to be free of air bubbles, which introduce a lag in the response of the system to changes in perfusion rate. The cannula, about 10 cm in length, can be more easily mounted and positioned if it is first bent in two places over a low flame. A 45° bend about 2–3 cm from the tip permits the micromanipulator to be advanced at a less oblique angle than with a straight cannula, and a dog-legged bend near the opposite end provides sufficient space between the cannula and microelectrode clamp to accommodate the diameter of the flexible tubing.

2. Rapid Flushing of the Tubule Lumen

The perfusion technique can be modified to permit rapid changing of the contents of the cannula. Such a modification has been used in a study of solute movements across the tubules by paracellular and transcellular routes (O'Donnell and Maddrell, 1983; O'Donnell *et al.*, 1983). The external bathing solution and the perfusion fluid contain equal concentrations of the radiolabeled solute of interest. When equilibration of solute concentration in the tissue is complete, the primary route through which the solute moves into the tubule lumen can be determined from the level of radioactivity in the tissue. A large number of counts per minute indicates equilibration of solutes throughout the cells, and therefore a transcellular permeation route. A small number of counts per minute suggests that the solute does not penetrate the cell membranes, and that its movement is restricted to a paracellular route through the lateral intercellular spaces. These measurements, however, require that the luminal contents of the tubule be rapidly flushed with "cold" Ringer's solution so that the number of counts in the lumen is negligible.

The contents of the cannula are first replaced by removing the flexible tubing, inserting a polyethylene tube filled with "cold" Ringer's solution, and advancing the tube as close to the tip of the cannula as possible. The polyethylene tube is first pulled to a fine tip (100 μm diameter) over a low flame. Once positioned, the polyethylene tube is connected to a syringe and the contents of the cannula are flushed out and replaced by the "cold" Ringer's. A low concentration of vital dye such as amaranth red is added to the Ringer's so that the progress of the flushing can be observed. The flexible tubing is then reconnected, and the syringe plunger is advanced manually so that the luminal contents of the tubule are flushed out in less than 2 sec. Observation of the dye in the "cold" Ringer's is again used to determine when the flushing is complete. Once flushed, the tubule is immediately pulled off the glass rod and pulled through two to three washes of "cold" Ringer's so as to remove counts adhering to the external surfaces of the tubule. The number of counts is then determined by liquid scintillation spectrometry.

Using this technique, we have found that small uncharged solutes such as xylose move primarily by a transcellular route, whereas large or charged molecules such as inulin are restricted to a paracellular pathway. A complete description of a series of such experiments is forthcoming (O'Donnell and Maddrell, 1983; O'Donnell *et al.*, 1983).

3. Simultaneous Perfusion of Internal and External Tubule Surfaces

We have recently developed a method for continuously perfusing both the luminal (apical) and basal surfaces of a Malpighian tubule. Our experiments were devised to permit measurement of osmotic fluxes of water across the tubule; our external Ringer's solution was therefore of lower osmotic concentration than the luminal perfusate (O'Donnell, *et al.*, 1982). We believe, however, that the technique may have general application whenever perfusion of both surfaces of a tubular tissue is desired.

The apparatus is shown in Fig. 2.2. The internal perfusion method has been described above. The external perfusion system consists of two stainless steel tubes that are formed by cutting hypodermic needles at right angles and then reaming the opening. The needles are attached to syringe barrels that are mounted on micromanipulators. Flexible tubing connects the syringe barrels to the inflow and outflow reservoirs of Ringer's solution. The steel tubes of the needles are bent so that their tips are parallel to the horizontal surface of the

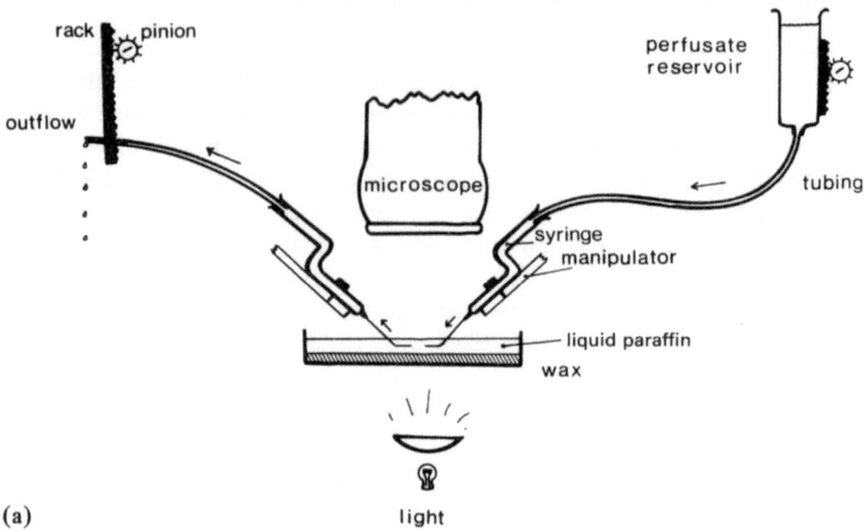

(a)

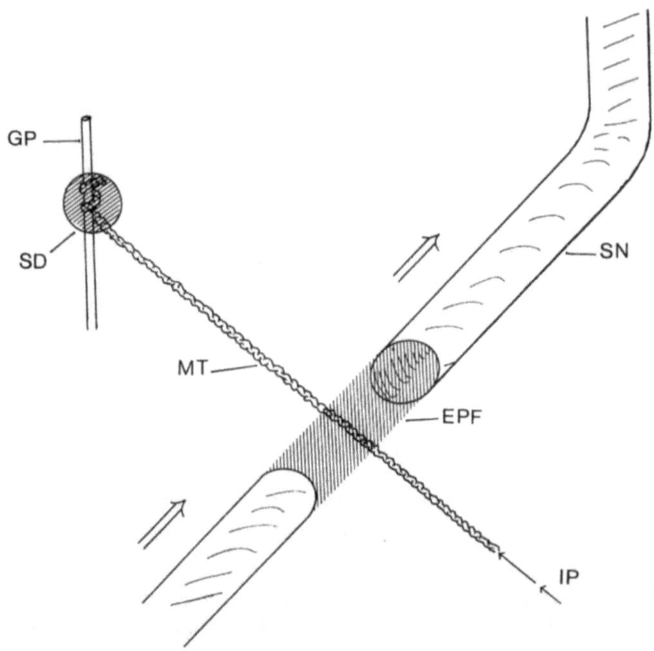

(b)

Fig. 2.2. Apparatus used for perfusing the external surfaces of a length of Malpighian tubule. **(a)** General view. **(b)** Close-up of the column of external perfusion fluid (EPF) connecting the syringe needles (SN). Osmotic water movements across the externally perfused region of the Malpighian tubule (MT) are determined by measuring the change in the rate of secretion from the tubule in the presence or absence of external perfusion. Rates are determined from the change in volume of the secreted droplet (SD), calculated from the droplet's diameter. The open end of the Malpighian tubule and the secreted droplet are held by surface tension to a glass pin (GP) embedded in a wax bed on the bottom of the petri dish. The direction of flow of the internal perfusate (IP) is indicated by the solid arrow. The external perfusate moves in the direction of the open arrows.

petri dish, with the syringe barrels and micromanipulator arms at 45° from the horizontal. The open ends of the steel tubes are placed about 1 mm apart.

During perfusion, fluid from a reservoir flows through the input tube, and the system is primed by sucking fluid into the outflow tube. Each reservoir is mounted on a rack and pinion so that its height above the petri dish can be adjusted. Once the system is primed, the flow rate and the shape of the fluid column between the tubes is adjusted precisely by altering the height of one or both of the reservoirs. The pressure head is opposed by the surface tension of the exposed column of perfusate at its interface with the paraffin oil.

This system permits very rapid exchange of the fluid in contact with the external surfaces of a length of Malpighian tubule. External perfusion rates of 20 $\mu l\ min^{-1}$ were typical. The radius of the fluid column surrounding the exposed length of tubule was generally 0.45 mm, so that flow rates over the external surfaces of the tissues were about $20/[(0.45)^2 \times 3.14] = 31\ mm\ min^{-1}$. The diameter of *Rhodnius* Malpighian tubules is about 0.05 mm, so that the fluid in contact with the basal surfaces is exchanged about 31/0.05 = 620 times per minute, or 10 times per second. Unstirred layer effects will, of course, somewhat decrease this rate of exchange.

IV. Measurement of Potentials

1. Transepithelial Potentials

Studies of ion transport often require the measure of transepithelial potential differences because ions can move down gradients in concentration or electrical potential. Transepithelial potential differences can be measured very simply in Malpighian tubule preparations maintained under paraffin oil. These preparations have been described previously (Maddrell, 1980a). Salt bridges, consisting of fine polyethylene tubing filled with 3 *M* KCl in 3-4% agar, are brought into contact with the bathing medium and secreted droplets, respectively, by means of micromanipulators. The potential differences are measured by connecting the salt bridges through calomel half-cell electrodes to an electrometer such as a Keithley Model 602. Any asymmetry potentials are determined by placing both salt bridges in the bathing droplet, and subtracting the result from the transepithelial measurements.

Transepithelial potential differences can also be readily measured during continuous perfusion of the bathing fluids. We have found that the easiest method is a modification of the technique described by Berridge and Prince (1972) and Prince and Berridge (1972) for the study of the salivary glands of the blowfly *Calliphora erythrocephala*. The experimental chamber used is shown in Fig. 2.3a and c. Two lateral compartments are separated by a paraffin-filled central compartment. Ringer's solution is perfused through one of the lateral compartments; the other lateral compartment is filled with Ringer's, but not perfused. The open end of the dissected tubule that has been placed in the perfusion compartment is

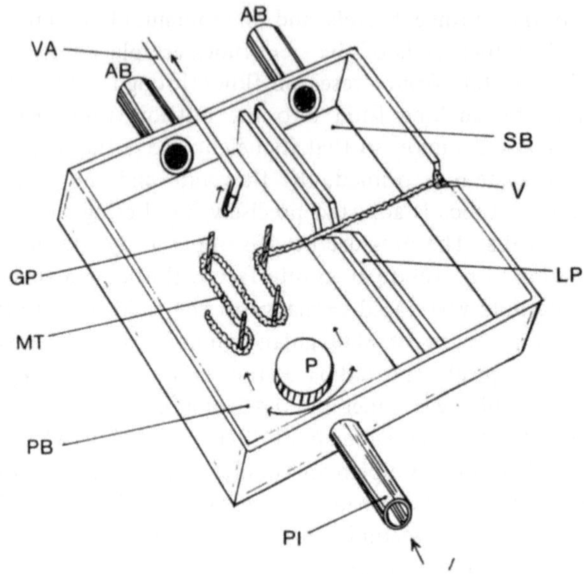
AB
VA
AB
SB
V
GP
LP
MT
P
PB
PI

(a)

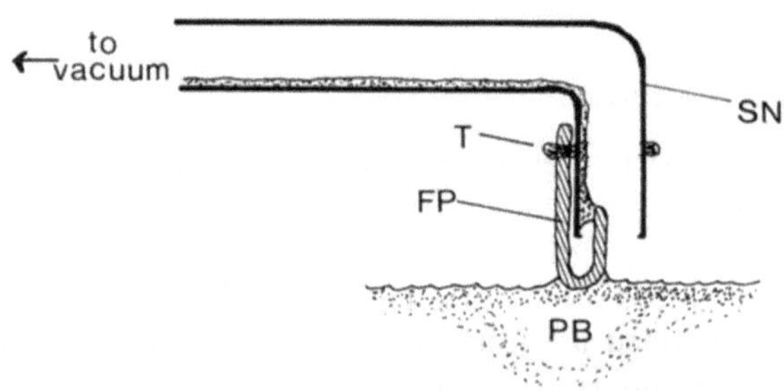
to
vacuum
SN
T
FP
PB

(b)

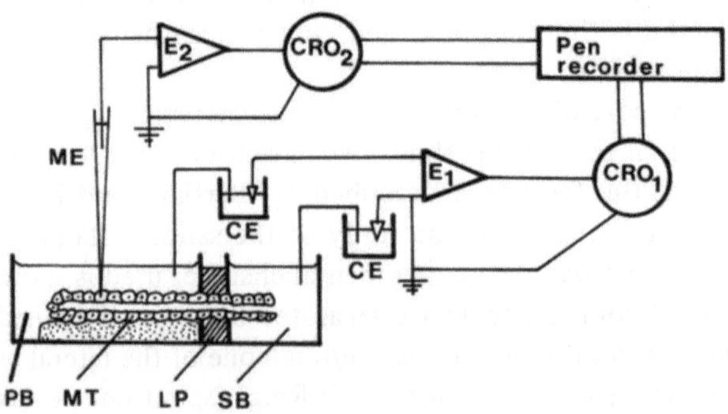
E2
CRO2
Pen
recorder
ME
CE
E1
CRO1
CE
PB
MT
LP
SB

(c)

pulled through the paraffin to the other lateral compartment. The tubule is secured by being tied with silk thread, and the end of the thread is stuck to the chamber with tacky wax. Alternatively, the open end of the tubule is stuck into a drop of silicone grease or petroleum jelly (Vaseline) in a groove in the outside rim of the chamber. The tubule wall is then nicked with a fine pair of scissors or a sharp glass sliver so that the luminal contents are electrically continuous with the bath.

The perfusion reservoirs consist of as many as eight conical flasks with stopcocks, mounted adjacent to the chamber and connected by flexible tubing to a manifold (Holder and Sattelle, 1972); the single outflow tube of the manifold is connected to the inlet of the perfusion chamber. The outflow port of the chamber is connected to a constant head device from which fluid is removed by a glass pipette connected to a vacuum aspirator. Alternatively, fluid can be aspirated through a wick that is placed either in the constant head device or in the perfusion chamber. To make the wick, a hypodermic syringe needle is cut at right angles and partially blocked with a strip of filter paper (Fig. 2.3b). The needle is connected to a vacuum aspirator through a syringe barrel and vacuum tubing. A vacuum flask is interposed between the other end of the tubing and the sink so as to prevent intermittent electrical grounding of the perfusion bath. The syringe barrel is mounted on a micromanipulator and the needle is positioned until the fluid in the perfusion chamber contacts the filter paper. The paper acts as a wick through which perfusion fluid is aspirated, and the level of

---

Fig. 2.3. Apparatus used for measuring transepithelial and intracellular potential differences. **(a)** Perfusion chamber. Fluid enters the perfusion bath (PB) through a perfusion inlet (PI). The bath is kept mixed by turbulence as fluid flows around the pillar (P) in front of the perfusion inlet. For the measurement of intracellular potential the Malpighian tubule (MT) is secured by being wrapped around a series of glass pins (GP) that are stuck onto a bed of agar on the bottom of a perfusion bath. The arrangement of the tubule as shown facilitates microelectrode penetration of cells along most of the length of the tubule. The potential difference between the perfusion bath and the secretion bath (SB) is the transepithelial potential difference, and is measured through agar bridges (AB), one in each bath. LP, liquid paraffin; V, Vaseline; VA, vacuum aspirator. **(b)** Device for continuously removing fluid from the perfusion bath (PB). A strip of filter paper (FP) is tied by a thread (T) to a bent syringe needle (SN), which is shown in longitudinal section. Once the filter paper contacts the perfusion bath, a continuous stream of fluid (stippled) is drawn through the paper and along the inside of the syringe needle. **(c)** Circuit diagram of the system used for simultaneous measurement of intracellular and transepithelial potentials. Explanation in text. CE, calomel electrodes; E_1, Keithley electrometer; CRO_1, CRO_2, cathode ray oscilloscopes; ME, microelectrode; E_2, high impedance amplifier suitable for intracellular recording. [From Prince, W.T. and Berridge, M.J. (1972). The effects of 5-hydroxytryptamine and cyclic AMP on the potential profile across isolated salivary glands. J. Exp. Biol. 56, 323–333.]

fluid in the chamber can be controlled by altering the height of the needle above the chamber surface. With no further adjustment, this device will maintain a constant level of fluid in the chamber over a wide range of flow rates. It is also unaffected by any air bubbles in the inflowing fluid, which tend to block the outflow tube and cause the chamber to overflow if the constant head device is used.

Transepithelial potential differences are measured through agar bridges placed in the lateral baths (Fig. 2.3c). The bridges are connected through calomel half-cells (CE) to an electrometer (E_1) and a cathode ray oscilloscope (CRO_1). The oscilloscope is connected in parallel with a two-channel pen recorder.

2. Measurement of Intracellular Potentials During Continuous Perfusion

Membrane potentials are measured through a microelectrode (ME in Fig. 2.3c) connected to a high impedance amplifier (E_2). The potential is displayed on a second cathode ray oscilloscope (CRO_2) and recorded on the second channel of the pen recorder. The potential is measured with reference to earth potential and, because either reference electrode in the two lateral baths could be earthed, the potential can be measured across the basal membrane with reference to the perfusion bath (PB) or across the apical membrane with reference to the secretion bath (SB).

To measure intracellular potential differences, a necessary modification of the system used for measuring transepithelial potential differences is the provision of a means for securing the tubule so that it is completely stationary during perfusion. This is accomplished by filling the perfusion chamber with 1 mm agar (stippled in Fig. 2.3c) and then looping the tubule around three or four glass pins stuck into the bed of agar (Fig. 2.3a). Once the tubule is secured, a conventional microelectrode holder mounted on a micromanipulator can be used to insert conventional electrolyte-filled glass microelectrodes. We find it easiest to penetrate the cell membranes if the Malpighian tubules are first stimulated to secrete by the addition of 5-hydroxytryptamine to the perfusion fluid. Stimulated tubules are swollen and more turgid than unstimulated ones, and can be more readily impaled by the microelectrode tip.

Cell impalement is also facilitated if the electrode is positioned 15-20 degrees off the vertical, relative to the cell surface. The electrode tip is then advanced against the tubule wall until the cell surface is visibly dimpled, as observed through a dissecting microscope. Alternatively, a stimulator can be used to pulse a square wave through the electrode. As the electrode is advanced towards the cell, a sharp, 20-50% increase in the height of the wave, as observed on an oscilloscope, indicates that the electrode tip has been blocked by contacting the cell surface.

At this stage, penetration of the cell can be accomplished using either of two methods. If the apparatus is mounted on a metal base plate, then gently tapping the plate or the base of the micromanipulator with the nail of the index finger produces vibrations that generally cause the electrode to 'pop' into the cell. An alternative method can be used if the microelectrode amplifier is fitted with capacitance compensation. Rapidly turning the capacitance compensation control to full scale for a fraction of a second causes the amplifier to oscillate. The oscillation acts as a microcautery at the electrode tip, and often results in a successful impalement.

Several criteria are used to judge the success of the impalement. The membrane potential should change rapidly, usually reaching 90% of its final value within a fraction of a second and then increasing to a stable value within a few more seconds. Potentials that vary by no more than a millivolt over the course of a minute indicate a successful penetration. In *Rhodnius* Malpighian tubules it is quite feasible to record stable membrane potentials for as long as an hour.

References

Berridge MJ, Prince WT (1972) Transepithelial potential changes during stimulation of isolated salivary glands with 5-hydroxy tryptamine and cyclic AMP. J Exp Biol **56**:139–153

Holder RED, Sattelle DB (1972) A multiway non-return valve for use in physiological experiments. J Physiol (Lond) **226**:2P–3P

Maddrell SHP (1964) Excretion in the blood-sucking bug, *Rhodnius prolixus* Stal. III. The control of the release of the diuretic hormone. J Exp Biol **41**:459–472

Maddrell SHP (1980a) Bioassay of diuretic hormone in *Rhodnius*. In: Miller TA (ed) *Neurohormonal techniques in insects.* Springer, New York: pp 81–90

Maddrell SHP (1980b) The functional design of the insect excretory system. J Exp Biol **90**:1–15

Maddrell SHP, Klunsuwan S (1973) Fluid secretion by in vitro preparations of the Malpighian tubules of the desert locus, *Schistocerca gregaria.* J Insect Physiol **19**:1369–1376

Maddrell SHP, Pilcher DEM, Gardiner BOC (1969) Stimulatory effect of 5-hydroxy tryptamine (serotonin) on secretion by Malpighian tubules of insects. Nature **222**:784–785

Maddrell SHP, Pilcher DEM, Gardiner BOC (1971) Pharmacology of the Malphigian tubules of *Rhodnius* and *Carausius*: The structure activity relationship of tryptamine analogues and the role of cyclic AMP. J Exp Biol **54**:779–804

Nicolson SW (1976) Diuresis of the cabbage white butterfly, *Pieris brassicae*; Fluid secretion by the Malpighian tubules. J Insect Physiol **22**:1347–1356

O'Donnell MJ, Aldis GK, Maddrell, SHP (1982) Measurements of osmotic permeability in the Malpighian tubules of an insect, *Rhodnius prolixus* Stäl. *Proc R Soc Lond* B **216**:267–277

O'Donnell MJ, Maddrell SHP (1983) Paracellular and transcellular routes for water and solute movements across insect epithelia. J Exp Biol (in press)

O'Donnell MJ, Maddrell SHP, Gardiner BOC (1983) Passage of solutes through the walls of the Malpighian tubules of *Rhodnius* by paracellular and transcellular routes. (submitted)

Phillips, JE, Maddrell SHP (1974) Active transport of magnesium by the Malpighian tubules of the larvae of the mosquito, *Aedes campestris.* J Exp Biol **61**:761–771

Prince WT, Berridge MJ (1972) The effects of 5-Hydroxytryptamine and cyclic AMP on the potential profile across isolated salivary glands. J Exp Biol **56**:323–333

Ramsay JA (1954) Active transport of water by the Malpighian tubules of the stick insect, *Dixippus morosus* (Orthoptera, Phasmidae). J Exp Biol **31**:104–113

Chapter 3

Methods for the Study of Transport and Control in Insect Hindgut

J. W. Hanrahan, J. Meredith, J. E. Phillips, and D. Brandys

I. Introduction

The hindgut plays a central role in renal function and osmoregulation in most insects. This organ selectively reabsorbs solutes and water from "primary urine" which is secreted into the gut lumen by Malpighian tubules. The rectum of the desert locust *Schistocerca gregaria* has been studied in some detail in an attempt to understand the physiology of excretion and to learn more about insect ion transport mechanisms at the cellular level (Hanrahan 1982; Phillips 1981). Sodium, potassium, chloride, water, amino acids, phosphate, and acetate are all reabsorbed from the rectal lumen into the hemocoel by active mechanisms (reviewed by Phillips 1980, 1981). A variety of preparations have been used in these studies, each having particular strengths and weaknesses.

The equipment and preparations that have been useful in studying ion transport across the locust rectum are described in this chapter, and the advantages and limitations inherent in these techniques are discussed. The reader is guided to appropriate literature where further methodological information may be obtained. The *in vitro* methods described in this chapter should be adaptable, with only minor changes, to other insect epithelia that can be cut to form a fairly large "flat sheet," notably the integument and other parts of the gut. For insects possessing a hindgut that is too small to prepare as a flat sheet, perfusion techniques normally applied to vertebrate renal tubules are more appropriate (Boulpaep and Giebisch 1978, Chonko et al. 1978; Dantzler 1977; Giebisch 1972, 1977; Gottschalk and Lassiter 1973; Ullrich et al. 1969; Windhager 1968). A perfused preparation has been used by Prusch (1974, 1976) to study transepithelial potential and isotopic fluxes across hindgut of the horsefly maggot

Sarcophaga bullata, and by Dow (1981b) to study the absorption of ions and water in locust midgut.

II. Salines

Artificial salines that mimic hemolymph have generally been used for measuring transport across the hindgut *in vitro,* despite the fact that solutes and water are normally absorbed *in vivo* from a high-K, low-Na Malpighian tubule fluid that is quite unlike hemolymph. Artificial hemolymph may be suitable for initial studies designed to demonstrate active transport; however, experiments should be repeated using a saline that resembles Malpighian tubule fluid in order to obtain quantitative information that is more relevant to an understanding of the transport process *in vivo.* Some experimental methods described in this chapter require that both sides of the tissue be bathed in identical solutions (as when measuring short-circuit current; Sect. IV.2). Rectal Cl absorption in locusts is apparently more dependent on K and metabolic substrates on the mucosal (luminal) side rather than serosal (hemolymph) side; therefore, it is more logical to study absorption using salines that resemble the luminal contents (Chamberlin 1981; Hanrahan 1982; see also below).

Miller (1979) listed salines that have been used in studies of cockroach nerve and muscle preparations, although these are generally somewhat different from normal hemolymph composition. If no realistic saline exists for a particular insect, it may be possible to design one based on literature values for the composition of hemolymph (Buck 1953; Florkin and Jeuniaux 1974; Stobbart and Shaw 1974; Wyatt 1961) and Malpighian tubule fluid (Maddrell 1971, 1977; Stobbart and Shaw 1974). In the absence of such data, hemolymph and Malpighian tubule fluids must be measured.

Hemolymph is usually obtained by puncturing the body wall. Small samples are best collected under mineral oil using micropipettes. These are often stored inside the micropipette at room temperature under water-saturated oil (Ullrich et al. 1969). However, Lechene and Warner (1979) found that although no concentration changes occurred in droplets that were ejected from the micropipettes and stored under oil for 58 days at -80°C, cation concentration did increase by 20% after 3 h in 8-nl droplets and increased 100% in fluid droplets having volumes of 127 pl when stored at room temperature under water-saturated oil.

Commercial microcapillaries are available for measuring quantities of 1 μl or more (e.g., Drummond Microcaps, Broomall, Pa.). For nano- and picoliter volumes, standard-volume micropipettes may be constructed and calibrated (Bonventre et al. 1980; Little 1974; Prager et al. 1965; Quinton 1976; Riegel 1970; Ullrich et al. 1969). To reduce adhesion of samples and protein to the glass surface, pipettes may be siliconized by immersion into a 0.1% v/v solution of silicon oil in acetone (1107 Dow Corning, Midland, Mich.) followed by baking on a hot plate ($> 300^\circ$C) for 15 min or, alternatively, by rinsing with Siliclad

(Clay Adams, N.Y.) or Prosil-28 (P.C.R. Research Chemicals, Gainesville, Fla.) and air dried.

Little (1977) and Greger et al. (1978) reviewed microanalysis methods, including the now standard methods of flame photometry (see Fletcher 1978), helium glow photometry, and atomic absorption spectrophotometry. Ehrlich and Diamond (1978) listed the advantages of flameless over flame spectrophotometers: in the flameless mode, sample volume is reduced from 1 ml after dilution to 2.5 μl or less, the entire sample is utilized, and atoms have longer residence times in the beam. They also described several techniques for improving sensitivity to Ca, Mg, Na, and K in flameless spectrophotometers. The Varian Model 1200 with carbon rod atomizer Model 90 was similar in sensitivity to the instrument made by Instrumentation Laboratories, but 30-150 times more sensitive to lithium than those made by Perkin-Elmer and Hitachi (Ehrlich and Diamond 1978). An alternative approach, electron probe microanalysis of fluid droplets, has the advantage that more than four elements can be quantified in one very small sample (typically 50-500 pl). Unlike atomic absorption, with this approach there is no problem of dilution errors (see reviews by Bonventre et al. 1980; Garland et al. 1978; Quinton, 1978; Rick et al. 1977; Roinei 1975). The composition of extracellular fluid is also obtained by using the electron probe in frozen-hydrated sections (Gupta and Hall 1979, 1981; Gupta et al. 1977; Moreton 1981).

Inorganic phosphate is found at high concentrations in the hemolymph of *Carausius morosus* (40 m*M*; Duchâteau et al. 1953; see review by Florkin and Jeuniaux 1974) and in Malpighian tubule fluid of locust (15 m*M*; Speight 1967; also see Maddrell 1977). Speight (1967) and Andrusiak (1974; Andrusiak et al. 1980) measured phosphate in rectal tissue and absorbate by the method of Ernster et al. (1950) and found it satisfactory. Sutcliffe (1962) used the method of King (1932) and Levenbook (1950) used the method of Allen (1940) to determine hemolymph phosphate.

For reasons of interference, sulfate is more difficult to measure in hemolymph than inorganic cations, and it is therefore rarely attempted. Roach (1963) adapted the conductometric method of Paulson (1953) to measure sulfate concentration in slug hemolymph (also see Little 1977). Swaroop (1973) and Burns et al. (1974) described sensitive spectrophotometric methods that might be applicable to insect fluids. Although tracer studies have shown that sulfate is not transported in locust rectum (Hanrahan 1982) and seems to be unimportant in preparing salines, it should be noted that addition of sulfate to salines lowers the activity coefficient of calcium, and this might have indirect effects in some insect tissues.

For large insects, the pH of individual or pooled samples ($>$ 60 μl) may be measured using a conventional "micro"-pH electrode (PHM 71, Radiometer, Copenhagen). An internal capillary electrode for determining pH in nanoliter volumes was developed by Khuri et al. (1967) and subsequently modified by Uhlich et al. (1968), Levine (1972), and Karlmark et al. (1971). Caflisch and

Carter (1974) and Sohtell and Karlmark (1976) designed microelectrodes for measuring CO_2 tension (PCO_2). Maffly (1968) described a conductometric method for total CO_2 (10–600 nmol). Little (1974) used a microdiffusion method for measuring total CO_2 in volumes of 1–6 nl. Vurek et al. (1975) developed a micromethod based on the heat produced when CO_2 reacts with a crystal of lithium hydroxide. This device is sold commercially as a Picapnotherm (Microanalytic Instruments, Bethesda, Md.). Karlmark (1973) and Karlmark and Sohtell (1973) described techniques for measuring titratable acid, ammonium, and bicarbonate in nanoliter samples. Karlmark et al. (1982) recently improved the methods by using a glass-type pH electrode rather than an antimony electrode, which is subject to interference (Caflisch et al. 1978; Green and Giebisch 1974; Malnic and Vieira 1972; Puschett and Zurbach 1974; Quehenberger 1977), and by incorporating a new titration circuit.

Malpighian tubule fluid is more difficult to sample than hemolymph, especially in dehydrated insects when fluid is secreted at very low rates *in vivo* (for *in vitro* methods of collecting fluid, see Maddrell 1980; Ramsay 1954). Ligating the gut makes it possible to obtain sufficient tubule fluid *in vivo* from one locust to analyze all major ions and amino acids (see Table 3.1). Furthermore, this fluid is probably secreted under more natural conditions than fluid collected using earlier *in vitro* methods.

Briefly, locusts are anesthetized by exposure to CO_2 for 3–4 min, and a lateral incision is made in the third and fourth abdominal segments. The ileum is held with a hooked glass rod while silk ligatures are tied immediately anterior and posterior to the proctodeal valve, where Malpighian tubules join the gut. A few drops of molten Tackiwax (Cenco, Central Science Co., Chicago, Ill.) are used to seal the incision. Insect gut is extraordinarily extensible (Hodge 1939), with the result that the space between ligatures becomes distended with fluid after several hours and may be sampled easily by puncturing. The K concentration of tubule fluid collected in this way (Hanrahan 1982) is very similar to that obtained *in vivo* by cannulation (Phillips 1964c) or obtained without ligation of the gut (Dow 1981a). Moreover, collecting tubule fluid over this period does not measurably alter the hemolymph (Hanrahan, unpublished observations). Positive hydrostatic pressure in the ligated section does not seem to affect the composition of tubular secretion and may actually reduce errors due to the contamination of tubule fluid by gut contents. The technique is further improved (Chamberlin 1981) by placing crystals of amaranth in the hemocoel before closing the incision. The dye is actively concentrated in the tubule fluid (Lee 1961; Mordue 1969) and serves as a check against leakage through the ligatures. A comparison of the composition of Malpighian tubule fluid collected by this method with that of hemolymph from *Schistocerca gregaria* is presented in Table 3.1.

Salines that have been used to bathe locust tissues during *in vitro* transport studies are listed in Table 3.2 (see Burton 1975 for a discussion of salines that have been used with nerve and muscle preparations). Sucrose is not metabolized by locust rectum (Chamberlin 1981) and is routinely added to adjust osmotic concentrations. Methyl sulfate has been used to compensate for the large anion deficit observed in Malpighian tubule fluid (Table 3.1; Hanrahan 1982). Gluconate, methanesulfonate, and methyl sulfate may all be used to replace Cl during flux experiments in Cl-free saline; however, gluconate or methanesulfonate are

Table 3.1. Composition of Body Fluids (m*M*) Collected from *Schistocerca gregaria* Forskål *In Vivo*[a]

Constituent	Hemolymph	Malpighian tubule fluid[b]
Na^+	103	47
K^+	12	165
Mg^{2+}	12	20
Ca^{2+}	9	7
Cl^-	107	88
HCO_3^-	13	NM[c]
Phosphate	6	12
Alanine	1.0	1.0
Aspartate	0.1–0.9	0.5
Asparagine	1.0	0.0
Arginine	1.5	0.0
Glutamate	0.1–1.0	0.8
Glutamine	4	0.5
Glycine	14	4.0
Histidine	1.4	0.0
Isoleucine	0.4	0.0
Leucine	0.4	0.0
Lysine	1.0	0.0
Methionine	0.4	0.0
Phenylalanine	0.7	0.0
Proline	13	38
Serine	2–4	1.0
Threonine	0.5	0.0
Tyrosine	1.0	0.0
Valine	0.6	0.0
Glucose	2.5	4.6
Trehalose	20	NM
Acetate	2–9	4
Citrate	2	NM
Malate	<0.1	NM
pH	7.1	>7.0

[a]Data from Chamberlin and Phillips (1982), Baumeister et al. (1981), Hanrahan (1982), and Speight (1967).

[b]Collected by gut ligation *in situ.*

[c]NM: not measured.

Table 3.2. Composition of Physiological Salines (m*M*) Used to Bathe Locust Excretory Systems *In Vitro*

	Simple salines[a]				Complex salines[b]			High K^+ salines[c]	
Constituent	A	B	C	D	A	B	C	A	B
NaCl	168	98	185	129	24.5	100	100	50	87.2
NaH_2PO_4	6.1	7		4.3			1.9		6.1
Na_2HPO_4							3.1		
$NaCH_3SO_4$							25		
$NaHCO_3$	2.1	22	24	10.2	10.5	10			2.1
KCl	6.4	20	11	8.6	8.5				87.2
KH_2PO_4			1						
K_2SO_4						5	5	5	
KCH_3SO_4								140	
$MgCl_2$		2		8.5	13				3.6
$MgSO_4$			2.5			10	10	10	
$CaCl_2$	3.6	2	3	2	2	5	1		2.2
Choline Cl								10	
Na_2 succinate					7.4				
Na_3 citrate		3.4			1.9				
Malic acid		2.8			12.8				
Alanine						2.9	2	2.9	
Arginine						1.0		1.0	
Asparagine						1.3		1.3	
Na glutamate		5	3		12.3				
Glutamine					2.6	5	5	5	
Glycine					2.7	11.4	15	11.4	
Histidine						1.4	2	1.4	
Lysine						1.4		1.4	
Proline					4.6	13.1	18	13.1	
Serine						1.5	2	1.5	
Tyrosine						1.9		1.9	
Valine						1.8		1.8	
Glucose	16.7	10	10	34	16.6	10	1	10	16.7
Trehalose							20		
Sucrose					80	100	49		+
Maltose					5.6				
Penicillin + streptomycin					+				
pH	7.0	–	7.0	7.2	6.7	7.1	7.4	7.1	6.9

[a] A: Goh and Phillips (1978); Mordue (1969). B: Maddrell and Klunsuwan (1974). C: Spring and Phillips (1980a, 1980b). D: Anstee et al. (1979).
[b] A: Williams et al. (1978), after Berridge (1966). B: Hanrahan (1982). C: Chamberlin (1981).
[c] Artificial Malpighian tubule fluid. A: Hanrahan (1982). B: Phillips et al. (1982).

preferred as a Cl substitute since they do not interfere significantly with Cl-sensitive, liquid ion-exchanger microelectrodes, permitting the same saline to be used for both tracer flux and microelectrode work (Hanrahan 1982). Although *N*-methyl-*D*-glucamine salts must be prepared from free base, *N*-methyl-*D*-glucamine is superior to choline as a sodium replacement during short-circuit current experiments since it is not sensed by K ion-exchanger resins and does not cause cell membrane depolarization (Duffey et al. 1978; Hanrahan 1982; Reuss and Grady 1979). It is important to use a permeant replacement ion when substitutions are made under open-circuit conditions; otherwise active transport may be rate limited by low counter-ion permeability.

Several other problems concerning physiological salines deserve special mention: Although many insects closely regulate the composition of hemolymph, some variation does occur and may be significant physiologically. Lettau et al. (1977) found diurnal variations in hemolymph K activity of approximately 50%. Furthermore, Pichon (1970) reported that hemolymph ion concentrations vary depending upon where in the hemocoel samples are taken. Clearly, the physiological state of an insect and the sampling site should be considered when analyzing body fluids and when designing salines for *in vitro* experiments.

With the exception of ion-sensitive electrodes, analytical methods measure total ion concentration rather than the more physiologically important quantity, ion activity. Treherne et al. (1975) measured the apparent Na activity coefficient (γ'_{Na}; ratio of activity to total concentration) in hemolymph of cockroaches (0.64) and locusts (0.73). Lettau et al. (1977) reported large diurnal oscillations in the apparent activity coefficient for K (γ'_K) in *Leucophaea* hemolymph and suggested that this might result from fluctuations in K binding to hemolymph proteins. Wiedler and Sieck (1977) used ultrafiltration to measure the fraction of ions that associate with macromolecules (molecular weight > 12000 daltons) in cockroach hemolymph and found that Na (22%), Mg (25.5%), Ca (16.2%), Cl (10.3%), and PO_4 (26.9%) are all bound significantly, whereas K ions are not. Florkin and Jeuniaux (1974) reviewed early literature on ion binding in insect hemolymph. Based on published stability constants for Mg- and Ca-glutamate complexes, Clements and May (1974) calculated that 48% of total hemolymph Ca and 73% of total Mg may be bound to amino acids in locust hemolymph. If a simple saline lacking organic acids and macromolecules is preferred over a more complex saline, then, ideally, the concentration of salts should be adjusted so that ion activities are similar to those in complete hemolymph or Malpighian tubule fluid.

Finally, appropriate metabolic substrates must be included in salines in order to maintain normal transport activity *in vitro.* For example, the rate of cyclic adenosine 3′, 5′-monophosphate (cAMP)-stimulated Cl transport across locust rectum is fivefold higher *in vitro* when bathed with saline containing 10 m*M* glucose and physiological levels of the 11 major amino acids in locust hemolymph, as compared to saline containing only 10 m*M* glucose as the energy source (Hanrahan 1982; cited by Phillips 1980). Moreover, recta exhibit a four-fold higher backflux of ^{36}Cl and electrical conductance after prolonged exposure to amino acid-free salines. These observations indicate deterioration and greater leakiness in the absence of amino acids (Hanrahan 1982). Chamberlin (1981) showed that proline is the substrate preferred by isolated mitochondria from locust recta, and proline is the most abundant substrate in rectal tissue (> 60 m*M*). This amino acid is also by far the most effective exogenous substrate in restoring active Cl transport across substrate-depleted tissue. Proline only causes maximum stimulation when added to the mucosal (lumen-facing) side. This is of considerable significance because proline is actively secreted into Malpighian tubule fluid (concentration 38 m*M*; Chamberlin and Phillips 1980, 1982). Clearly, if specific substrate requirements are unknown, any saline that is designed to maintain transport activity should include a full range of metabolic substrates that are normally available both in the gut lumen and in the hemolymph.

III. *In Situ* Preparations

1. Injection and Retrieval from Ligated Recta

The transport of salts and water across insect hindgut was first studied quantitatively in the locust rectum by injection and retrieval of fluid from ligated locust recta *in vivo* (Phillips 1964a, 1964b, 1964c).

A. Method

The apparatus used in this method is shown in Fig. 3.1a. (The notation used in Fig. 3.1a is indicated within parentheses in the description given below.)

Locusts are starved 3 days to reduce the amount of food in the gut and are anesthetized by exposure to CO_2 (3 min) followed by ether (3-5 min). An incision is made through the cuticle of the sixth abdominal segment, and silk thread or human hair is looped around the ileum. A loose ligature is tied, moved posteriorly to the anterior margin of the rectal pads, and then tightened (1); care is taken not to damage the large tracheae that supply the epithelium. After the incision is sealed with beeswax-resin mixture, locusts are held immobile in a small wire cage fastened to the stage of a dissecting microscope (2). A glass injection needle that tapers to 0.5 mm o.d. (3) is attached to one arm (a) of a three-way stopcock (4), inserted into the ligated rectum, and sealed into place with beeswax-resin so as to occlude the anus (5). The second arm (b) of the stopcock is connected to a micrometer burette (6). This arrangement allows injection of fluid (40 μl) into the rectal lumen when arm (c) of the stopcock is closed.

Rectal fluid is sampled at timed intervals (30-45 min) by inserting a glass capillary (or polyethylene tubing) through arm (c). To measure fluid absorption rate, radio-iodinated serum albumin (alternatively, ^{14}C-inulin or ^{14}C-polyethylene glycol) is included in the injection fluid to serve as a volume marker. Albumin is excellent for this purpose since it neither crosses the rectal cuticle at signifi-

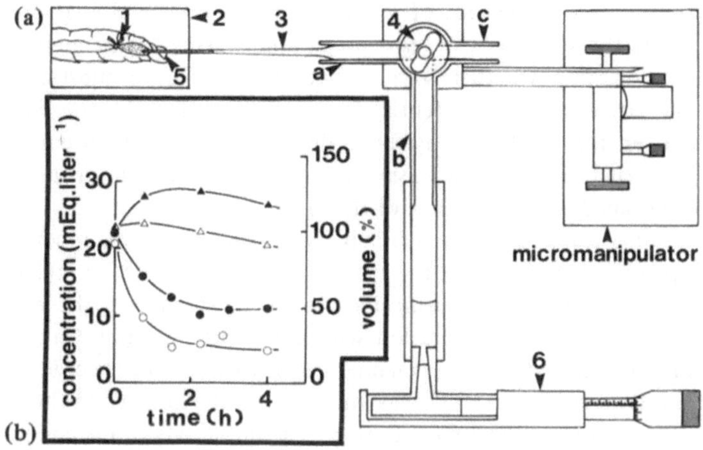

Fig. 3.1. Measurement of net absorption from ligated locust rectum *in situ*. **(a)** Apparatus. See text for description. **(b)** Typical results for changes in volume (triangles; original volume = 100%) and Cl concentration (circles) of rectal contents with time after a hyperosmotic sugar solution of low NaCl content is injected into the lumen. Open and closed symbols represent two different preparations. [From Phillips JE (1964b) Rectal absorption in the desert locust *Schistocerca gregaria* Forskål. II. Sodium, potassium and chloride. J Exp Biol **41**:39–67]

cant rates, nor is it adsorbed onto the surface of the cuticular intima lining the rectal lumen (Phillips 1964a). Moreover, albumin is completely soluble (at low concentration) over the wide range of pH and osmotic pressures present *in vivo* in the rectum. To measure ion uptake rates, samples of rectal fluid are analyzed by standard flame photometry methods for Na and K, and Cl is determined by electrometric titration with $AgNO_3$ (Ramsay et al. 1955). Drag effects due to fluid movement can be controlled by adjusting luminal osmolality with sucrose if so desired. The required concentration of sucrose must be determined empirically for each species.

Examples of the time course of changes in volume and Cl concentration are shown in Fig. 3.1b. Net ion absorption rates are calculated from changes in ion concentrations and volume of luminal fluid, as described by Phillips (1964a, 1964b). This method served to demonstrate active Cl transport in locust rectum and also showed that water is absorbed for several hours from a hyperosmotic solution of xylose, indicating that fluid absorption is not strongly dependent on net transepithelial ion absorption.

B. Advantages and Limitations

With this technique, net uptake of salts and water is measured under conditions that closely resemble those in unperturbed insects: minimal surgery is required and the epithelium remains well oxygenated because tracheae are left intact. It is important to have some idea of *in vivo* transport rates and characteristics if one is to assess the success of *in vitro* preparations. This *in situ* method is adequate to demonstrate whether some active transport of a substance occurs, provided that transepithelial potential is also measured under the same conditions; moreover, this method is useful to demonstrate maximum concentration gradients that such active mechanisms can generate across the rectal wall (i.e., because of the small luminal volume of fluid).

Unidirectional fluxes are very difficult to measure using this *in vivo* preparation. Phillips (1964b) estimated the rate of sodium backflux by injecting ^{24}Na into the hemolymph and measuring its rate of appearance in the rectal fluid after allowing for tracer equilibration with the locust Na pool. Chloride backflux could not be estimated in this manner because of the low specific activity of ^{36}Cl. Instead, fluid containing ^{36}Cl was injected into the lumen and then sampled for both Cl concentration and radioactivity. The initial rate of decline in specific activity is then a rough estimate of the rate at which unlabeled Cl enters the rectal fluid from the hemolymph.

The injection-and-retrieval method relies on changes in the composition or volume of rectal fluid and consequently does not permit measurement of steady-state transport rates. After fluid is injected, neither the chemical nor electrical gradients between hemolymph and rectal fluid can be controlled by the investigator. Furthermore, these gradients change during the course of the experiment as the composition of rectal fluid is altered by reabsorptive processes. It is difficult to separate out quantitatively active and passive components of the total net absorption when chemical and electrical gradients are present and changing. For example, with this technique, some active K absorption was indicated for locust rectum and was presumed to represent a high proportion of total K absorption. More recent *in vitro* studies, however, have clearly shown that most K ($>$ 80%) is absorbed passively by electrical coupling to active Cl transport (Hanrahan and Phillips 1982) and the active K component is relatively small.

Like all *in situ* techniques, the injection-and-retrieval method does not isolate the rectum from natural neural or hormonal factors that might influence transport rates *in vivo*. These probably vary with time and between individual insects even under well-controlled rearing conditions. This characteristic may be an advantage or disadvantage depending on the purpose of the experiment. Unknown neural or hormonal factors complicate studies of cellular transport mechanisms; nevertheless, regulation of hindgut function and its physiological significance can ultimately be resolved only through some use of an *in situ* preparation.

2. *In Situ* Perfusion of Locust Recta

The *in situ* perfusion method has been used to study the ionic dependencies of transepithelial potential under steady-state conditions in intact locusts–information that could not be obtained using the injection-and-retrieval approach

(Hanrahan, in preparation). A somewhat similar perfusion method has been used *in vitro* with radioactively labeled molecules to measure the permeability properties of gut cuticle (Phillips and Beaumont 1971; Phillips and Dockrill 1968; Maddrell and Gardiner 1980).

A. Method

The apparatus used in this method is shown in Fig. 3.2a. (The notation used in Fig. 3.2a is indicated within parentheses in the description given below.)

Perfusion cannulae (1) are prepared in advance from polyethylene tubing (PE 10, Dickinson and Co., Parsippany, N.J.) as follows: Quick-setting epoxy resin is mixed with dye (Sudan Black) and drawn into the tubing a distance of 5 mm by suction. After hardening, the black epoxy plug (2) is trimmed to 3 mm and a small hole (3) is made about 2 mm from the plug. Large cannulae (4) are made from PE 90 tubing cut into 3-cm lengths and one end is flared by means of a flame. To make agar bridges (5), 3-5 g agar (Bacto-agar, Difco Laboratories, Detroit, Mich.) is dissolved in 100 ml hot 3 *M* KCl and drawn by suction into PE 10 tubing, where it solidifies. Transepithelial potential (V_t) is

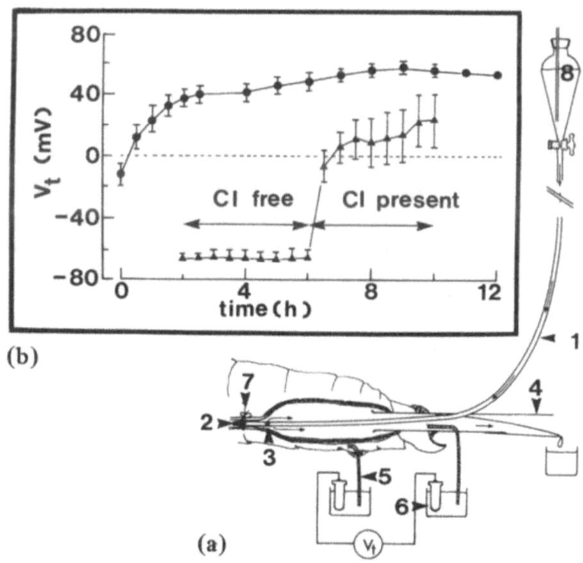

Fig. 3.2. Measurement of transrectal potential difference (V_t) while perfusing the lumen of ligated locust recta *in situ*. **(a)** Apparatus. See text for description. **(b)** The V_t measured *in situ* by this method when locust recta are perfused with control saline (100 m*M* Cl, circles), or with Cl-free saline (triangles) and subsequent readdition of Cl to control levels at the fifth hour. The sign of V_t refers to the lumen relative to the hemocoel (mean ±SE). J Hanrahan, unpublished observations.

measured by using a differential electrometer (616, Keithley Instruments, Cleveland, Ohio) with high input impedance, connected by standard calomel electrodes (6) that are immersed in small pots of 3 *M* KCl. The V_t is recorded at slow speed on a strip chart recorder (B-5000 Omniscribe, Houston Instruments, Austin Tex.).

Locusts are anesthetized (see above) and fastened to a plasticine block. A large cannula (4) is inserted into the rectum and secured externally with tissue glue. After a small incision is made on the sixth abdominal segment, the ileum is held gently with a hooked glass rod while the blocked end of the PE 10 cannula is passed through the larger cannula (4) to the posterior ileum. The perfusion cannula is ligatured (7) in place at its plugged end (the plug is visible through the gut wall as a black dot in the lumen). Perfusate is delivered from a reservoir (8) set up to maintain a constant pressure head throughout the experiment. Perfusion rate is controlled by adjusting the height of the reservoir. In locusts, transepithelial potential is not dependent on perfusion rate over the range 100–170 μl/min (maximum volume of the rectum is 40 μl).

The data shown in Fig. 3.2b obtained during *in situ* perfusion with normal artificial Malpighian tubule fluid (AMTF, i.e., high-K saline A; Table 3.2). As locusts recover from the operation, V_t (lumen relative to hemocoel) increases from negative to strongly positive values (+51.5 mV), and then remains constant for at least 12 h. When locust recta are perfused with Cl-free AMTF, V_t remains negative (−65.8 mV) until Cl is restored. These *in vivo* results are consistent with active electrogenic Cl absorption as demonstrated by *in vitro* methods (Hanrahan and Phillips 1982c; Spring and Phillips 1980c; Williams et al. 1978). Under anoxia, V_t declines to near zero when Cl^- is present, which provides further evidence for active CL transport *in vivo* (Hanrahan, unpublished observation). Note that V_t approaches the transepithelial equilibrium potential for K (E_K = −56 mV) during anoxia and during perfusion with Cl-free AMTF. This *in vivo* result suggests that transepithelial K movements are largely passive, as predicted by *in vitro* studies (Hanrahan 1982), and that an important function of electrogenic Cl absorption is to drive passive K absorption by making V_t more positive than E_K.

B. Advantages and Limitations

This method allows measurement of V_t in intact locusts under steady-state conditions. The composition of the rectal fluid is maintained constant by perfusion of the rectum at a high rate. Rectal Cl absorption is controlled by a neurosecretory peptide in locusts (Phillips et al. 1980; Spring and Phillips 1980a, 1980b; Spring et al. 1978), and there is evidence that feeding normally causes release of this factor into the hemolymph (Hanrahan 1978; Phillips et al. 1981; Spring and Phillips 1980c). Using this preparation, it may be possible to follow hindgut Cl transport activity *in vivo* during dehydration, feeding, the diurnal cycle, etc., and thereby detect when Chloride Transport-Stimulating Hormone (CTSH) is normally released and by what sensory trigger. This is feasible because locusts survive for several days when cannulated in this way and they recover completely when ligatures and cannulae are subsequently removed. However, it must be appreciated that lack of change in transepithelial V_t does not exclude stimulation of Cl transport by CTSH if permeability to the counter-ion K increases proportionally (this could be checked by measuring the effects of transient ion substitutions on V_t). Since ionic fluxes are not measured directly, it is essential to verify which transport processes generate V_t, for example, using *in vitro* tracer methods (see Sect. IV.2). Net transport has been estimated *in vivo* by measuring the current necessary to clamp V_t at 0 mV (Küppers and Thurm 1980); however, this is only a valid measure of "short-circuit" current under nonphysiological conditions, i.e., when the luminal perfusate resembles hemolymph rather than Malphigian tubule fluid. The surgery required for *in situ* perfusion is mild compared to many other insect preparations (Miller 1979); however, it remains to be shown that normal hormonal (or neural) control of rectal transport processes are unperturbed by this method.

IV. *In Vitro* Preparations

1. Everted Rectal Sacs

This everted rectal sac preparation has been used to study the hormonal control (Mordue 1969; Phillips et al. 1980) and ionic requirements of fluid transport (Goh and Phillips 1978; Phillips et al. 1982) and also the transport of amino acids (Balshin and Phillips 1971), phosphate (Andrusiak et al. 1980), and acetate (Baumeister et al. 1981) across locust rectum. It has also been adapted for measuring unidirectional tracer influxes across the apical cell border (Hanrahan 1982; Hanrahan and Phillips 1982).

A. Method

The apparatus used in this method is shown in Fig. 3.3a and b. (The notation used in Fig. 3.3a and b is indicated within parentheses in the description given below.)

After the head, legs, and wings are removed, a locust is pinned onto a plasticine block and a lateral incision is made in the posterior three segments. After the flap of cuticle is pinned back (1), a flared PE 90 cannula (2) (see also Sect. III.2.A) is inserted through the anus, pushed beyond the anterior margin of the rectal pads, and tied with a silk ligature at (3). The ileum is then cut at (4), all tracheal connections are severed, and the rectum is gently everted by withdrawal of the cannula until the posterior margin of the rectal pad emerges. The everted cylinder is freed by a transverse cut at (5), between the posterior edge of the pads and the anus, and the gut is rinsed by injection of 1 ml saline through the

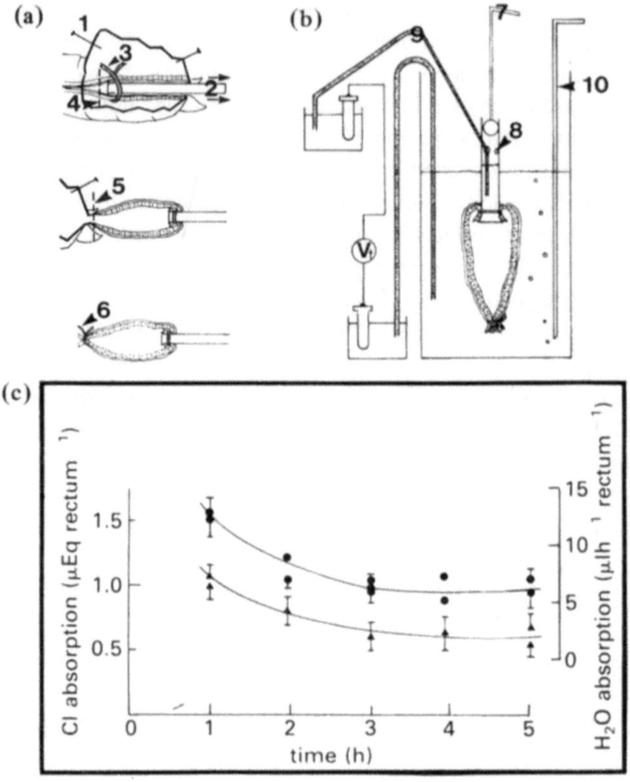

Fig. 3.3. Measurement of absorption and transrectal potential (V_t) across cannulated, everted rectal sacs. See text for description. **(a)** Method for preparing everted rectal sacs. **(b)** Apparatus for measuring absorption and recording V_t. **(c)** Typical time courses of water (circles) and Cl (triangles) absorption rates for unstimulated everted rectal sacs bathed in simple saline A of Table 3.2. [From Goh S, Phillips JE (1978) Dependence of prolonged water absorption by *in vitro* locust rectum on ion transport. J Exp Biol **72**:25–41]

cannula. A second ligature is tied at (6), and any fluid remaining in the sac is removed by suction through the large cannula via PE 10 tubing attached to a syringe. A standardized volume of saline (< 50 μl) is then injected into the sac through the cannula.

A bent insect pin (7) supports the sac during incubation and weighing and reduces evaporation of the absorbed fluid. Small holes near the top of the cannula (8) prevent air pressure from building up during fluid absorption and provide access for 3 *M* KCl agar bridges (9), which are used to monitor V_t. External saline is vigorously stirred with 95% O_2/5% CO_2 delivered through PE tubing (10), except in HCO_3-free salines, when 100% O_2 is used (we have observed no difference in Cl transport using these two gas mixtures; Hanrahan and Phillips 1980a, 1982). To measure fluid absorption, sacs are removed from the saline bath, blotted dry by being lightly touched with absorbant tissues, and weighed at intervals with an accuracy of ±0.25 mg. Fluid removal is aided by the hydrophobic surface of the cuticle. Absorbate inside the sacs is removed and analyzed by standard methods (Sect. II) and then replaced with a known volume of fresh saline. Alternatively, no saline is added and absorbate is allowed to collect undiluted over each hourly incubation period.

The time course of Cl and water absorption in everted rectal sacs is shown in Fig. 3.3c (Goh and Phillips 1978; for further results see Phillips et al. 1982). Both Cl and fluid absorption rate decline especially during the first hour and reach a near steady state. Even after 5 h *in vitro,* recta absorb 5 μl fluid per hour and the mucosal side is +12 mV with respect to the serosal side, quite similar to values measured *in situ* (Goh and Phillips 1978).

To measure the rate of Na and Cl uptake across the apical cell border (Hanrahan 1982), everted sacs are placed for 20 min in control saline containing ^{3}H-mannitol, which serves as an extracellular space marker that can cross the rectal cuticle. To permit equilibration of tracers with the subintimal space (which is essentially a large unstirred layer), tissues are transferred to saline containing ^{3}H-mannitol and ^{36}Cl or ^{22}Na at pH 4.5 for 30–45 min. At this pH (which is within the normal physiological range) it is possible to inhibit active Cl transport temporarily while equilibrating tracers with the extracellular fluid. After timed exposure to a test solution containing ^{3}H-mannitol plus ^{36}Cl or ^{22}Na at pH 7.0, tissues are dissected off of the cannula, blotted on bibulous paper, weighed on tared pieces of aluminum foil to ±0.1 mg, macerated in vials containing 1 *N* KOH, and digested at 80°C overnight. After cooling, samples are neutralized with H_2SO_4 and counted by standard techniques: by a liquid scintillation counter for ^{3}H and ^{36}Cl, or by an automatic gamma counter for ^{22}Na. The radioactivity of ^{3}H-mannitol and ^{36}Cl (or ^{22}Na) are measured in triplicate in 1-μl samples. The number of ^{3}H-mannitol counts provides an estimate of extracellular space which is then used to correct for extracellular ^{36}Cl and ^{22}Na.

B. Advantages and Limitations

Viability is greatly enhanced when sacs are tied onto cannulae inside-out so that the mucosal (or luminal) side is bathed in a large volume of vigorously oxygenated saline. Instead of falling drastically to near zero within 1 h as in uneverted sacs, fluid transport and V_t decline and then approach a steady-state condition which is maintained for more than 5 h (Goh 1971; Goh and Phillips 1978; Phillips et al. 1982). Everted sacs are also supplied with metabolic substrates more efficiently because proline, which is the preferred substrate for supporting active Cl transport in locust rectum, is most stimulatory when added to the mucosal side (Chamberlin 1981), and there is an excess supply of substrates because of the large external volume compared to that inside the sacs.

Everted sacs are simple to prepare and provide the most convenient *in vitro* method for measuring the rate of net fluid absorption. The dependence of net fluid transport on ions, metabolic substrates, and inhibitors may be examined by adding them to the external saline. Effects of hormone extracts may be tested by injecting them into the sac via the cannula if the injected solution is replaced frequently to prevent dilution by the rectal absorbate (Mordue 1972; Phillips et al. 1982). Moreover, large changes in concentration of transported solutes may appear over short periods because of the small internal volume. This can be an advantage if the investigator wishes primarily to demonstrate net active transport against a net electrochemical gradient, especially if convenient radiotracers are not available for studies using Ussing chambers, or if considerable metabolism or dilution of the labeled compound with a large tissue pool of unlabeled compound is likely (e.g., phosphate; see Andrusiak et al. 1980).

This preparation is less well suited for studying ion transport mechanisms since unidirectional tracer fluxes across the rectum (and hence flux ratios) are more conveniently measured using a flat-sheet preparation. In rectal sacs, transepithelial potential is not controlled by the investigator, and ion and water absorbed by the sac tends to alter the composition of any fluid that is placed inside. Furthermore, accurate measurement of tracer influxes into the rectal tissue across the apical cell border is hampered by the need to preequilibrate tracers with the subintimal space at low pH. Since it is difficult to completely remove the intima from sacs without damaging the underlying epithelium, sacs will probably be of limited use in future studies of flux across the apical border, except for those insect species in which the intima is directly attached to the epithelium so that no subintimal space is present.

2. Short-Circuit Current/Tracer Fluxes

In this method, electrical and chemical gradients across the epithelium are abolished by clamping the transepithelial potential at 0 mV using current from an external source (i.e., "short-circuiting" the transepithelial potential) when identical solutions are placed on both sides of the epithelium. Under these conditions and in the absence of solvent drag, net flux of an ion must be due to active transport. The short-circuit current (I_{sc}) required to maintain V_t at 0 mV is the sum of active ion transport processes under these conditions. Cylindrical preparations have been used to measure both I_{sc} and tracer fluxes in lepidopteran midgut (Harvey and Nedergaard 1964) and I_{sc} alone in locust rectum (Herrera et al. 1976, 1977); however, most laboratories now use flat-sheet preparations, which ensure homogeneous current density through the tissue, are generally more convenient, and allow accurate correction for saline resistance. A practical guide to short-circuit current methods is given by Watlington et al. (1970).

A. Method

a. Chambers. The Ussing-type chamber was first adapted to measure tracer fluxes across silkworm midgut (Wood 1972; see Wood and Moreton 1978). Wood's design was modified by Williams (1976; see Williams et al. 1978) for use with locust rectum, and it is this modified version that is shown in Fig. 3.4a.

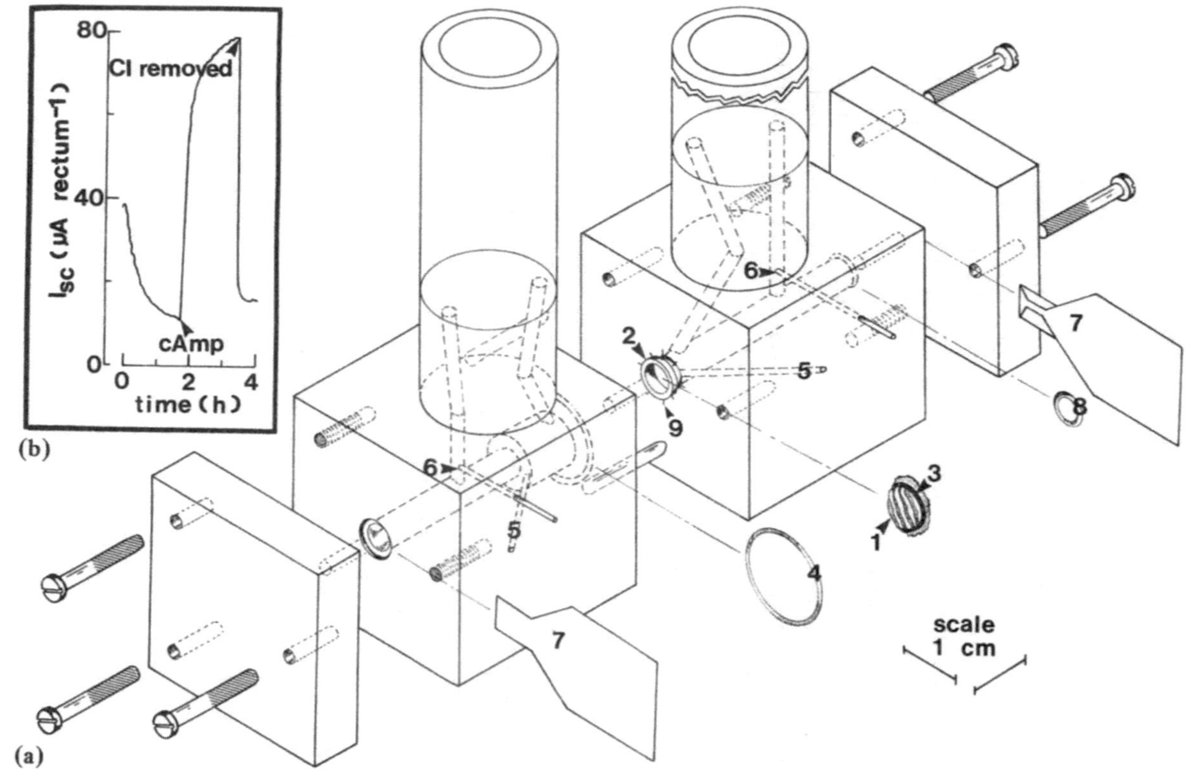

Fig. 3.4. Measurement of I_{sc}, V_t, and radiotracer fluxes across locust recta. **(a)** Isometric projection of Ussing chambers used by Williams et al. (1978). **(b)** Typical trace of I_{sc} measured across a locust rectum using these chambers, showing stimulation of I_{sc} by 1 m*M* cAMP and abolition of ΔI_{sc} when Cl is replaced by cyclamate or other anions. [From Hanrahan JW (1982) Cellular mechanism and regulation of KCl transport across an insect epithelium. Doctoral dissertation, University of British Columbia, Vancouver, Canada]

Briefly, rectal tissue (1) is mounted on a collar-shaped opening (2) and secured with a rubber O-ring (3). The half-chambers are clamped together in a vice-like frame. To measure V_t, 3 *M* KCl agar bridges (size PE 90) are placed near the tissue through ports (5). Each agar bridge connects with a reservoir of KCl and standard calomel electrodes as described before. Short-circuit current is passed between silver electrodes (7), which are sealed by means of O-rings (8) to either end of each half-chamber. Solutions are stirred vigorously with 100% O_2 or 95% O_2/5% CO_2, which enters each half-chamber (6) after being humidified by bubbling through a water trap to reduce evaporation from the reservoirs. A common problem is the replacement of tungsten wire pins (9) (0.005-inch diameter, 1-2 mm length) that project from the sides of the collar (2). Holes are bored half-way through the collar under a dissecting microscope by using an insect pin in a standard pin holder. A sharp cutting edge is achieved by cutting the insect pin at a slight angle with wire cutters. The collar is fixed to the chamber and the pins to the collar with a viscous glue made from ethylene dichloride containing Plexiglas shavings. This hardens in seconds and the seams are then sealed with epoxy (Araldite, Ciba-Geigy Plastics and Additives Co., Duxford, Cambridge, U.K.). Finally, we suggest sealing agar bridges into the chamber initially with silicone sealant (Dow Corning Canada, Mississauga, Ont.). They can later be removed, leaving a relatively tight-fitting opening for subsequent use.

b. Dissecting and mounting rectum. Locusts are pinned out as described in Sect. IV.1.A. Flaps of cuticle are pinned back, and the rectum is cut longitudinally by scissors inserted in the anus. Tracheal connections are severed and the rectum is removed as a flat sheet by transverse cuts that are made anterior and posterior to the rectal pads. To mount tissue on the collar-shaped opening, both the top and gas inlet of the chambers are plugged. The half-chamber is tilted horizontally and filled with oxygenated saline; a tissue is stretched flat over the opening by means of watchmakers forceps and hooked on fine tungsten pins (9). A rubber O-ring ($\frac{1}{4}$-inch i.d., $\frac{3}{8}$-inch o.d., $\frac{1}{16}$-inch Width) is trimmed to halve its thickness and the flat surface is placed around the end of a metal cylinder having a slightly larger diameter than the rectal tissue. The cylinder is placed lightly against the tissue and forceps are used to slide the O-ring over the tissue, thereby making a tight seal to the collar.

c. Electrical methods. A variety of circuits for short-circuiting epithelia have been published (e.g., Flemström et al. 1973; Isaacson et al. 1971; LaForce 1967; Rothe et al. 1969; Schoen and Candia 1978; Smith 1978; Wood and Moreton 1978; Wright 1967). Since insect epithelia generally have low electrical resistance, external saline contributes significantly to the voltage drop between agar bridges when a short-circuit current (I_{sc}) is passed (Rehm 1975; Wood and Moreton 1978). Noncompensating voltage clamps were used in early studies of locust rectum (Herrera et al. 1976, 1977; Spring and Phillips 1980a, 1980b, 1980c; Williams et al. 1978), dragonfly rectum (Leader and Green 1978), and silkworm midgut (Harvey and Nedergaard 1964). A compensating voltage clamp that employs three voltage-sensing electrodes has been used to short-circuit midgut of silkworm (Wood 1972; Wood and Harvey 1975, 1976; Wood and Moreton 1978), *Manduca sexta* (Blankemeyer 1977, 1978; Blankemeyer and Harvey 1978), *Rhodnius prolixus* (Farmer et al. 1981), and integument of silkworm (manually short-circuited; Jungreis and Harvey 1975) and *Manduca* (Cooper et al. 1980; Cornell and Jungreis 1981). The compensating voltage clamp designed by Rothe et al. (1969) has been updated for use in tracer flux and microelectrode studies of locust rectum (Fig. 3.5a–c; Hanrahan 1982; Hanrahan and Phillips 1980a, 1980b, 1982). Also, a simple dual-tracking power supply, suitable for use with a two-channel voltage clamp is shown in (Fig. 3.5d). The voltage clamp used in recent studies of locust rectum is described below (refer to Fig. 3.5b).

Briefly, transepithelial potential ($V_t = E_+ - E_-$) is amplified (gain = 100X) by a differential amplifier (A1) having high input impedance (4253, Teledyne Philbrick, Dedham, Mass.) and is monitored at (1). When in "run" mode, a second operational amplifier (A2; 725, National Semi-conductor Corp., Santa Clara, Calif.) passes current (I_{sc}) between I_{Hi} and I_{Lo} in order to bring ($E_+ - E_-$) to 0 mV. A third amplifier (A3; 308, Fairchild, Mountain View, Calif.) is set up in current-to-voltage configuration. I_{sc} is monitored at (2). Some fraction of this signal (set by potentiometer P3) is fed back to correct for series resistance of the saline.

In the description that follows, switches (SW) and potentiometers (P) are numbered as they appear on the front panel in Fig. 3.5a; their location in the circuit is shown in Fig. 3.5b. The numbers enclosed by ovals in Fig. 3.5b refer to the position of the rotary switch (SW9 in Fig. 3.5a and c), which is connected to a strip chart recorder (220, Soltec Corp., Sun Valley, Calif.) through "–" and "+" terminals (Fig. 3.5a and c). The recorder is set at 10 V full-scale deflection. During construction of the instrument, amplifier A1 is brought to 0 output by adjusting the 100-KΩ potentiometer P5 with input to A1 grounded (SW7 to E_{os}) and the amplifier isolated (SW1 to "Adj. 0 offset"). These switches (SW1 and SW7) are returned to their run positions, where they remain for operation of the voltage clamp. Similarly, SW2 and SW5 remain in the "on" position, and SW6 and SW8 in the "off" position ("*" for SW6) for routine operation.

To operate the voltage clamp, agar bridges are positioned in the chambers, which are assembled without a tissue present and filled with saline at constant temperature. Potential-sensing and current-passing electrodes are connected and grounded at the input on the front panel (V +, – for potential; IN, OUT for current; GRND), as are the recorder leads (RECORD +, –). Any asymmetry between the potential-sensing calomel electrodes (monitored in SW9 position 1) is nulled by using battery B_1 (SW2 in "on" position) and 20-KΩ potentiometer P1.

To set the automatic compensation for saline resistance, a mock I_{sc} is passed from the power supply (SW4 to "Adj. IR") through chambers containing only saline. This current is monitored in SW9 position 2, where an output of 100 mV is equivalent to 1 μA. The mock I_{sc} is adjusted to 80–90 μA by using the ADJ IR COMP potentiometer P2. With amplifier A2 at unity gain (SW3 to "test"), the ADJ IR COMP potentiometer P3 is set so that the output at SW9 position 3 remains at 0 mV when the I_{sc} is turned on and off (SW4 from "Adj. IR" to "*"). The numerical setting of P3 is noted; it is characteristic of the saline used and the position of the agar bridges. SW4 is returned to off ("*"), SW3 to "run," and SW9 to position 2, and a tissue is then mounted. Transepithelial potential (V_t) is monitored in SW9 position 1. To short-circuit the tissue, SW4 is switched to

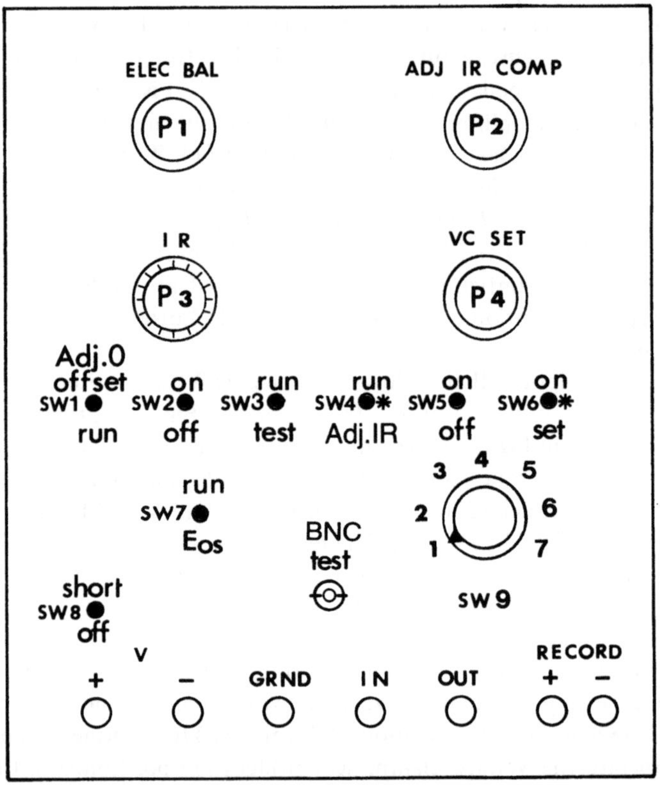

(a)

Fig. 3.5. Electrical apparatus to short-circuit locust recta with automatic correction for saline resistance. **(a)** Front panel of apparatus showing the position of switches (SW), potentiometer controls (P), and input–output terminals. **(b)** Circuit diagram indicating the orientation of the tissue (M, S) potential-sensing electrodes (E), and current-passing electrodes (I), as well as the location of the switches and potentiometers referred to in part a. See text for detailed description. **(c)** Relationship of the rotary switch (SW9), recorder terminals, and BNC connector in part a and the numbered points on the circuit diagram in part b. **(d)** Circuit diagram of the simple dual-tracking power supply used with a two-channel voltage-clamping device of the type shown in parts a–c. [From Hanrahan JW (1982) Cellular mechanism and regulation of KCl transport across an insect epithelium. Doctoral dissertation, University of British Columbia, Vancouver, Canada]

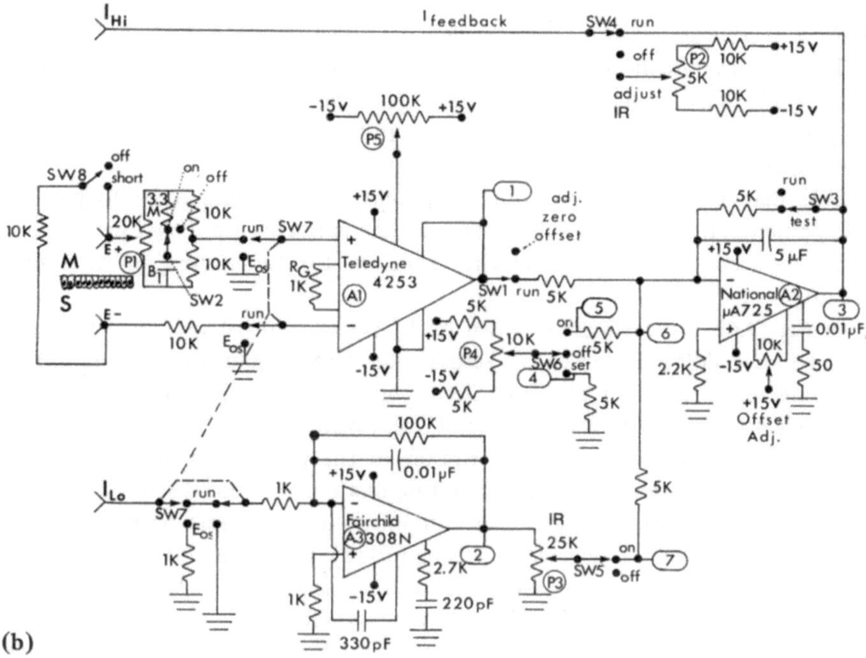
I Hi
I feedback
SW4 run
off
adjust IR
P2 10K 5K
10K
+15 V
-15 V
-15v 100K +15v
P5
SW8 off short
on off
3.3 M
20K
10K
10K
E+
P1
M
S
B1
SW2
run SW7
E os
R G 1K
Teledyne 4253
A1
+15V
-15V
1
adj. zero offset
SW1 run 5K
5K
+15V
P4
10K
-15V
5K
on 5
5K
off
SW6 set
4
6
5K
run
5K SW3
test
+15V
5 µF
National A2 µA725
3
0.01µF
2.2K
-15V
10K
50
+15V Offset Adj.
E-
10K
run
E os
100K
0.01µF
+15V
I Lo
run
SW7
1K
Fairchild A3 308N
E os
1K
1K
-15V
2.7K
2
220pF
330pF
IR
25K
P3
SW5
on
off
7
5K

(b)

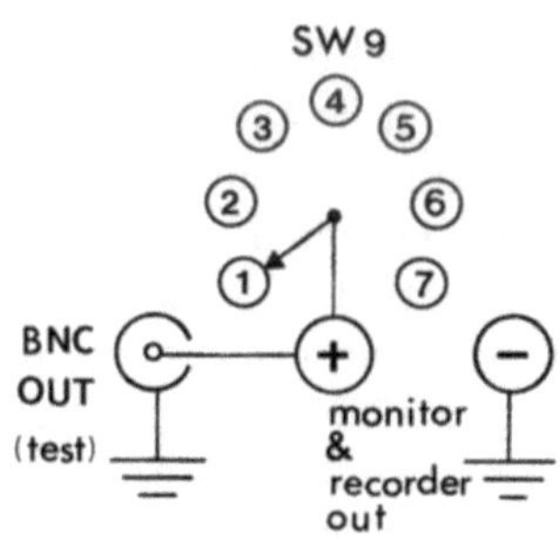
SW9
1
2
3
4
5
6
7
BNC OUT (test)
+
−
monitor & recorder out

(c)

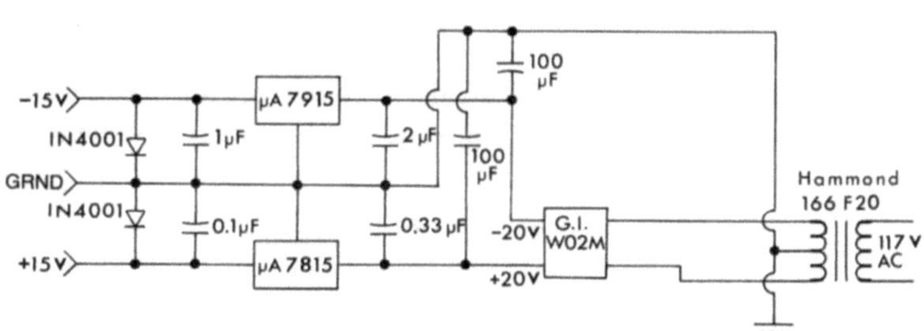
-15V
IN4001
GRND
IN4001
+15V
1µF
0.1µF
µA7915
µA7815
2µF
0.33µF
100 µF
100 µF
-20V
+20V
G.I. W02M
Hammond 166 F20
117 V AC

(d)

"run" and the current is monitored at SW9 position 2 (a 100-mV output equals 1 μA, with recorder set at 10-V full-scale deflection).

The V_t may also be clamped at some other voltage. With the output monitored at SW9 position 4 and SW6 at "set," clamping voltage is selected by adjusting potentiometer P4. When SW6 is moved to "on," the tissue is clamped at the new voltage, which can be monitored SW9 position 5.

SW8 may be used to short-out the potential-sensing electrodes to prevent development of asymmetry when the machine is not in use. Positions 6 and 7 of SW9 are convenient test points for checking the functioning of the clamp circuitry.

In Fig. 3.4b a recording is shown of I_{sc} across a locust rectum before and after 1 m*M* cAMP is added to the serosal side. Over 90% of the I_{sc} is abolished following replacement of external Cl with gluconate, confirming that Cl transport is electrogenic (Spring and Phillips 1980b; Williams et al. 1978).

d. Tracer methods. In this description of the protocol used in measuring tracer fluxes across locust rectum, counting methods are not discussed in detail since they depend on the particular equipment used. Sample preparation and counting techniques are reviewed by Peng (1977), Peng et al. (1980), and Wang et al. (1975); for physical data regarding isotopes, see Brodsky (1978).

After tissues have reached a steady-state condition at approximately 4 h (Hanrahan 1982; Williams et al. 1978), tracer is added to one side (termed "side 1"), and allowed to mix for several minutes. Aliquots are taken from side 1 with microcapillary pipettes (Drummond Microcaps, Broomall, Pa.) and counted. For liquid scintillation counting of ^{14}C and ^{3}H organics, ^{36}Cl, and $^{35}SO_4$, it is necessary to use xylene/surfactant-based or dioxane-based fluors such as ACS (Amersham Corp., Arlington Heights, Ill.), Aquafluor (New England Nuclear, Boston, Mass.) or Permablend (Packard, Warrenville, Ill.), which will accept 1–2 ml water. The number of counts per minute (cpm) is converted to disintegrations per minute after a "percentage efficiency" curve is constructed using commercial reference standards according to the channels ratio or external standard methods. If no reference standards are available for a particular isotope, cpm may be used in calculating flux rates if quenching (and hence percentage efficiency) is kept constant by the addition of identical amounts of saline to all samples. In this case, the "effective" specific activity of tracer on side 1 is determined by taking 1-μl aliquots. These aliquots are quenched with the same volume (1 ml) of "cold" saline as are samples obtained from side 2 during experiments.

Samples of saline from side 2 are replaced with unlabeled saline and residual radioactivity per unit volume is calculated to correct for this dilution. Unidirectional flux is estimated over each interval by the equation (adapted from Shaw 1955),

$$J_{1\to 2} = \frac{a_2 \cdot V \cdot C}{a_1 \cdot T \cdot A} \tag{3.1}$$

where $J_{1\to 2}$ is the unidirectional flux rate (moles per square centimeter per hour) a_1 is the concentration of radioactivity (in cpm per milliliter) on side 1, a_2 is the increase in concentration of radioactivity on side 2 during the flux period (in cpm per milliliter), V is the volume of saline on side 2 (in milliliters), C is the total concentration of bulk plus tracer isotope on side 1 (in moles per milliliter). T is the time interval over which flux occurred (in hours), and A is the area of the membrane. If tracer activity remains low on side 2 relative to side 1 (well below 5%), no correction is needed for tracer backflux from the "cold" to "hot" side. To compare I_{sc} and tracer fluxes quantitatively, it is necessary to integrate recordings of instantaneous I_{sc} over each flux sampling period. We use a planimeter for this purpose (Model L30M, Lasico, Los Angeles, Calif.)

Ideally, both influx and efflux should be determined simultaneously by using two isotopes (e.g., ^{22}Na and ^{24}Na), or separately on paired membranes. This is not generally feasible for insect epithelia. A solution to this problem is to use only preparations with similar electrical characteristics (e.g., I_{sc}, V_t, resistance) for measuring influx and efflux independently.

B. Advantages and Limitations

This approach allows the investigator to control both electrical and chemical gradients across the epithelium, and unidirectional ion fluxes may be accurately followed with time by using radiotracers. Active transport can be established unequivocally and quantified when the epithelium is bathed bilaterally with identical solutions and V_t is clamped at 0 mV (Ussing 1948, 1949; Ussing and Zerahn 1951). In the absence of solvent drag, net flux of a tracer under steady-state I_{sc} conditions measures the rate of active transport, although absence of net flux does not necessarily show that movements are passive (Rehm 1975). The preparation described here is ideal for characterizing the basic properties of transepithelial transport. For example, the transepithelial flux kinetics may be obtained by varying the external concentration of the transported ion with V_t at 0 mV. Coupled ion movements are tested by measuring tracer fluxes when other ions are replaced by large impermeant ions. The selectivity of active transport is measured directly using tracers of other potential substrates (e.g., see Zerahn 1980). Alternatively, when transport is electrogenic, other potential substrates may be substituted for the transported ion and their effects on I_{sc} noted. Exchange diffusion is indicated when unidirectional flux of a tracer requires the presence of the same ion on the opposite (or "trans") side.

In "leaky" epithelia, most transepithelial diffusion occurs via the paracellular pathway. Since this route constitutes a single barrier, epithelial permeability to ion i (P_i) may be calculated from the passive flux from side one to side two ($J^i_{1\to 2}$) under I_{sc} conditions according to

$$P_i = J^i_{1\to 2} c^i_1 \tag{3.2}$$

where c^i_1 is the concentration of ion i on side 1. This relationship may also hold for tight epithelia when the permeability of one cell membrane is much lower than that of the other.

The effects of specific inhibitors, putative hormones, and transmitters on active transport and epithelial permeability are readily investigated using this *in vitro* preparation because rates can be followed with time on the same preparation during experimental perturbations. In locusts, transrectal I_{sc} has been used as an assay system in the identification and purification of the new peptide hormone CTSH (Phillips et al. 1980; Spring et al. 1978). Finally, since both chemical and electrical gradients are controlled by the investigator, ion fluxes may be quantified under conditions that mimic those that exist *in vivo.*

The I_{sc}/tracer flux approach considers the epithelium as a "black box" and provides limited information concerning events at individual cell membranes (see Sect. IV.3). Mounting of insect epithelia on chambers may be complicated by small size or fragility. Deterioration prior to mounting is minimized by dissecting tissues on a cold plate. Mounting of tissues in Plexiglas chambers may lead to "edge" damage (see Helman and Miller 1973, 1974; Lewis and Diamond 1976; Walser 1970). The flux of $^{35}SO_4$ from serosa to mucosa has been used to estimate the maximum contribution of edge damage to total Cl permeability of locust recta mounted in the chambers shown in Fig. 3.4. By correcting for the relative free solution mobilities of Cl and SO_4, it was calculated that less than 12% of transepithelial ^{36}Cl backflux could be due to edge damage, and that this "leak" flux is less than 1.5% of the rate of active Cl transport (Hanrahan 1982). Small insect tissues are particularly susceptible to edge damage when mounted in chambers because of their larger edge-to-area ratios.

Although K and Cl each have only one convenient radioisotope (^{42}K, ^{36}Cl), unidirectional fluxes may, in some instances, be measured simultaneously using ^{86}Rb and ^{77}Br as tracers for K and Cl if membrane mechanisms do not discriminate between these isotopes. However, in locust rectum, the rate of Br absorption is less than one-half that of Cl and is therefore a poor substitute for ^{36}Cl. In *Cecropia* midgut, Rb is carried by the K pump with 60% efficiency, although this varies considerably between tissues (Zerahn, 1980). Carefully performed control experiments are essential when different radioisotopes are used as tracers. Some tissues apparently discriminate between different isotopes of a single element (termed "isotope interaction"; see DeSousa et al. 1971; Jacquez 1980; Li et al. 1974).

3. Intracellular Microelectrode Techniques

Once the mechanism of net flux has been characterized under steady-state conditions using tracers, the investigator is generally interested in measuring the electrochemical potential profile across the cell borders using ion-sensitive microelectrodes. If net flux occurs against an opposing transmembrane electrochemical gradient, an "active" transport mechanism is indicated at that membrane. Several methods that have been useful in studying KCl transport across locust rectum are described below (Hanrahan 1982; Hanrahan and Phillips 1982).

A. Ion-Sensitive Microelectrodes

a. Chambers. There are several considerations in designing a microelectrode chamber: efficient oxygenation and stirring, a similar (preferably identical) method of mounting tissues as was used in tracer flux studies (to permit legitimate comparisons), ready access of the epithelial surface to microelectrodes, and good visibility of the tissue and microelectrode during impalements. The Plexiglas chamber shown in Fig. 3.6a meets these criteria and also permits independent solution changes on both sides of the tissue without dislodging microelectrodes. Saline is delivered from well-oxygenated reservoirs (1) by gravity feed. The volume of saline in each half-chamber is adjusted by using a micromanipulator to raise and lower and suction syringes (2). Multiway valves (3) (Holder and Sattelle 1972) are placed near the chamber inlets for rapid switching between several different perfusates. A dissecting microscope (100X, Zeiss, Jena, D.D.R.) is used to observe the tissue and microelectrode through a glass window (4) at the mucosal end of the chamber. Fiber optics (Intralux 150H, Volpi AG, Urdorf, Switzerland) with twin arm illuminators (10008.001, Volpi AG) light the preparation from front and rear (5) through holes cut in the current-passing electrodes (6). Microelectrodes (16) are mounted in the micromanipulator (7) as shown in Fig. 3.6b. For stability, the chamber (8) is fastened to a solid Plexiglas stand (9), which in turn is bolted to a micromanipulator base (10; Leitz, Wetzler, F.R.G.) between two Leitz micromanipulators. This base rests on a heavy steel plate (11) (1.3 X 60 X 90 cm) supported by large rubber stoppers (size 15) (12). To further reduce vibrations, experiments are performed on a concrete table (13). The concrete table top (10 X 60 X 180 cm) rests on a 7-mm thick layer of lead (14) and a 15-mm layer of felt (15).

It is necessary to cut the cuticle to permit penetration of the rectal epithelial cells with microelectrodes from the lumen side. Fortunately, this cuticle is attached to the cell layer only at the edges of rectal pads in locusts and probably many other insects. It is relatively easy to lift up a fold of cuticle with fine forceps and cut it away from the rectal pad (Hanrahan 1982).

b. Fabrication and calibration of ion-sensitive microelectrodes. Capillary glass (1.0 mm o.d., Frederick Haer and Co., Brunswick, Me.) is cleaned for at least 2 h in concentrated HNO_3, thoroughly rinsed with distilled water, dried in an oven, and stored dust free at room temperature and 0% relative humidity (i.e., in a desiccator over H_2SO_4). Two glass capillaries are held together at both ends by pieces of silastic tubing (Sargent-Welch Sci., Weston, Ont., Canada) and mounted in a vertical microelectrode puller (PE-2, Narishige Scientific Instrument Laboratories, Tokyo), softened in the heating coil without any vertical displacement, rotated 180°, and then drawn to a final tip diameter of less than 1 μm. One barrel is then bent as shown in Fig. 3.6a (17) by heating over a small gas flame. After the reference barrel is filled with reagent-grade acetone (ACS, Eastman Kodak Co., Rochester, N.Y.), the electrode tip is dipped into a 0.1% solution (v/v) of Dow Corning 1107 silicone oil in acetone (1 μl/ml) for ap-

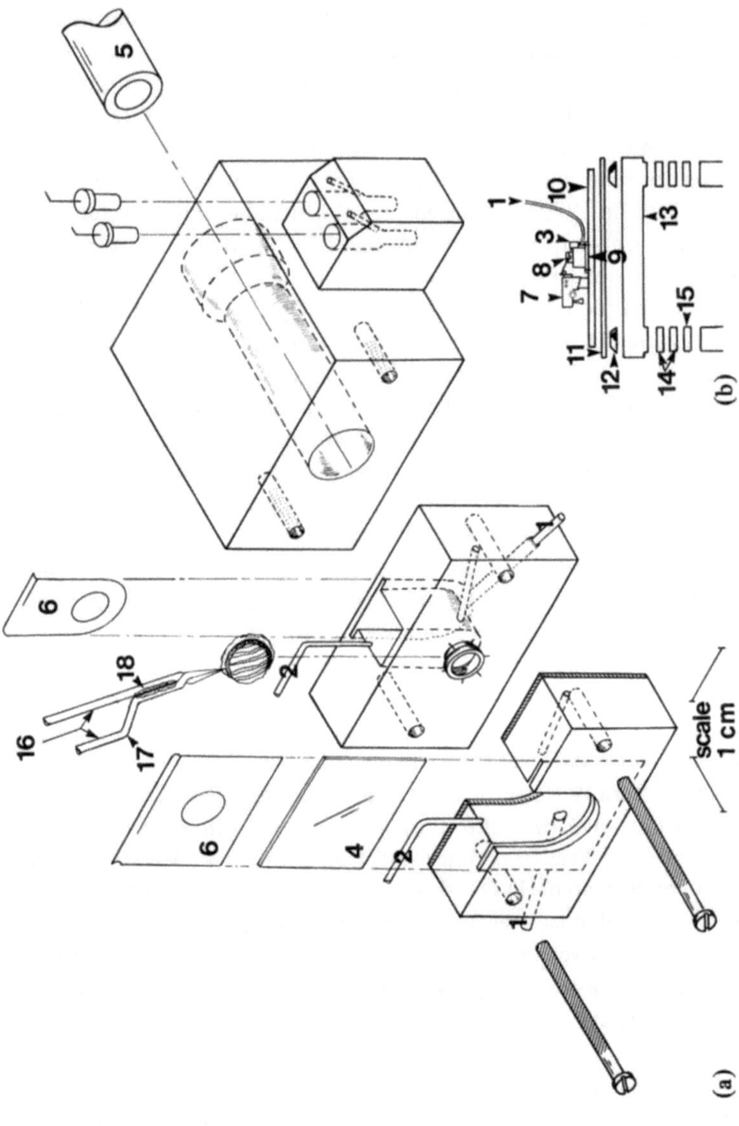

1
2
3
4
5
6
7
8
9
10
11
12
13
14
15
16
17
18
scale
1 cm
(a)
(b)

Table 3.3. Liquid Ion-Exchangers and Solutions Used in Fabricating Ion-Sensitive Double-Barreled Microelectrodes[a]

Electrode	Ion-sensitive barrel: Resin	Ion-sensitive barrel: Backing electrolyte	Reference barrel
K^+	Corning 477317 Corning Med. Prod., Medfield, Mass.	0.5 *M* KCl	1.0 *M* Na acetate
Cl^-	Orion 92-17102, Orion Res.; or the newer Corning 477913	0.5 *M* KCl	1.0 *M* Na acetate
Na^+	Monensin, 10% w/w in Corning 477317[b]	0.49 *M* NaCl at pH 3.0 (0.1 *M* citrate buffer)	0.5 *M* KCl

[a] From Hanrahan JW (1982) Cellular mechanism and regulation of KCl transport across an insect epithelium. Doctoral dissertation, University of British Columbia, Vancouver, Canada.
[b] See Kotera et al. (1979).

proximately 10 sec to coat the inner surface of the ion-sensitive barrel with a hydrophobic layer. It is essential that the acetone is kept dust and water free. Electrodes are then cured by laying them on glass microscope slides on a hot plate at 300°C for 15 min. Once the microelectrode shaft has been reinforced with fast-drying epoxy (18, Fig. 3.6a), microelectrodes are stored dust free in a desiccator over concentrated H_2SO_4 until needed.

A column of liquid ion-exchange resin (2-4 mm; Table 3.3) is injected into the silanized barrel from a syringe through fine polyethylene tubing made from PE 10 pulled out over a flame. Electrodes are usually constructed the day before experiments and allowed to fill by capillary action overnight. The microelec-

Fig. 3.6. Chambers used to mount recta as flat sheets for intracellular recordings with double-barreled, ion-sensitive microelectrodes. The chamber design is modified from that shown in Fig. 3.4a to permit viewing of the epithelium with a dissecting microscope and a fiber-optic light source (5) through glass-covered holes in the current-passing electrodes (6). Each chamber is perfused independently with oxygenated salines through inlets (1) and suction outlets (2). Transrectal potential (V_t) is recorded as in Fig. 3.4 and the preparation can be short-circuited, or current pulses can be passed across the rectal epithelium through the silver-foil electrodes (6). **(a)** Isometric projection of chambers. **(b)** Details for mounting this chamber to minimize vibrations. See text for a detailed description of recording methods. [From Hanrahan JW (1982) Cellular mechanism and regulation of KCl transport across an insect epithelium. Doctoral dissertation, University of British Columbia, Vancouver, Canada]

trode is then backfilled with the solutions listed in Table 3.3 and stored tip down in 0.5 *M* KCl (K and Cl electrodes) or 0.5 *M* NaCl (Na electrodes). A fine-tipped soldering iron brought close to the microelectrode glass surface as well as a cat whisker inserted in the electrode barrel are useful for coaxing resin to the tip and removing large bubbles. Immediately before use, microelectrodes are beveled by holding the tips in a fine jet stream of abrasive 0.05-μm gamma alumina (Micropolish, Buehler, Evanston, Ill.) suspended in distilled water (see Ogden et al. 1978) for 0.5-1 min. Excessive beveling should be avoided, particularly if electrodes are to be used in small cells, where leakage from the reference tip might raise intracellular K and Cl activities (Fromm and Schultz 1981; Lindemann 1975; Nelson et al. 1978).

Alternative methods for fabricating double-barreled liquid ion-exchanger microelectrodes have been described (Baumgarten 1981, Cl; Berridge and Schlue 1978, K; Fujimoto and Kubota 1976, K and Cl; Fujimoto et al. 1980, HCO_3; Khuri et al. 1972, K; Khuri et al. 1974, HCO_3; Kotera et al. 1979, Na; White 1977; reviewed by Khuri and Agulian 1981; Zeuthen 1980). Methods for constructing single-barreled ion-sensitive microelectrodes also vary (Brown 1976; Duffey et al. 1978; Garcia-Diaz and Armstrong 1980; Lee et al. 1980, Ca; Lewis and Wills 1980; Lewis et al. 1978; O'Doherty et al. 1979; Palmer and Civan 1977; Reuss and Weinman 1979; Spring and Kimura 1978; Walker 1971; Wills et al. 1979; reviewed by Thomas 1978).

Ideal ion-sensitive microelectrodes would respond according to the Nernst equation (i.e., 58.8 mV/decade change in ion activity at 22°C) when placed in different electrolyte solutions and would be perfectly selective for the ion of interest (or primary ion). Because microelectrodes do not have ideal properties, sensitivity and selectivity must be measured experimentally (Armstrong and Garcia-Diaz 1980; see the extended Nicolsky equation and problems associated with its use in Bindslev and Hansen 1981). Microelectrodes are calibrated frequently during the course of experiments in solutions that encompass the entire range of intracellular and extracellular ion activities. For example, 5, 50, 120, and 500 m*M* KCl solutions are used to calibrate microelectrodes for both K and Cl, except during perfusion with nominally K- or Cl-free salines, when 1 m*M* KCl is also included in the calibration series (Hanrahan 1982). Ion activities are calculated from the modified Debye-Hückel equation:

$$-\log \gamma = \frac{A\,|Z_1 Z_2|\sqrt{I}}{1 + \mathring{a}\,B\sqrt{I}} - 0.055\,I \tag{3.3}$$

where $I = \frac{1}{2}\,\Sigma\, c_i Z_i^2$, γ is the activity coefficient, A and B are Debye constants (Robinson and Stokes 1965), $\mathring{a}$ is the ion size parameter (Bates 1973), I is the ionic strength, c_i and Z_i are the concentration and valence of ion i, respectively.

The simplest method of calibrating microelectrodes is the separate solution method (Moody and Thomas 1971, method Ia; see also Fujimoto and Kubota 1976). When electrode electromotive force (EMF) is measured in separate solutions and plotted against the log activity of primary (a_i) and interfering (a_j) ions, the selectivity coefficient K_{ij} is the ratio a_i/a_j, giving identical microelectrode EMFs. However, this approach is only valid if calibration lines are linear and parallel in solutions containing primary and interfering ions. Selectivity coefficients are sometimes concentration dependent, especially with liquid ion-exchanger-type microelectrodes: in this case, selectivity is most usefully measured in solutions that resemble intracellular fluid. Selectivity coefficients may be determined using the fixed interference method (Moody and Thomas 1971, method IIa; Saunders and Brown 1977) by plotting microelectrode EMF against the activity of i when EMF is measured in the presence of a fixed amount of j (which should approximate that found intracellularly). Instead of selectivity coefficients being calculated by this method, experimentally measured activities may also be read directly from this "mixed solution" calibration curve (Garcia-Diaz and Armstrong 1980). Kotera et al. (1979) determined selectivity coefficients of Na-sensitive microelectrodes from the difference in electrode EMFs measured in two different calibrating solutions: one containing pure NaCl (120 m*M*) and the other, an isotonic mixture of NaCl (20 m*M*) and KCl (100 m*M*). Sodium-sensitive glass microelectrodes have higher selectivities than do liquid ion-exchanger types (see Thomas 1978, for details of constructing Na- and pH-sensitive glass microelectrodes).

Interference due to HCO_3 and other unidentified anions may result in artifactually high values of a^c_{Cl}, although this error is probably minimal when the perfusate is CO_2- and HCO_3-free (Bolton and Vaughan-Jones 1977; Brown 1976; Garcia-Diaz and Armstrong 1980; Saunders and Brown 1977). "Apparent" Cl activity, which is observed after several hours of perfusion with Cl-free saline, may be subtracted from measurements if it is assumed that cells are actually Cl free and that this error signal represents anion interference that is present under normal experimental conditions (Reuss and Grady 1979; Spring and Kimura 1978). Chloride-sensitive microelectrodes that are virtually insensitive to HCO_3 may be fabricated by fastening a silver wire into the open tip (1-2 μm diameter) of a glass microelectrode (Neild and Thomas 1973; Saunders and Brown 1977), or by first silver coating one (sealed) microelectrode and then enclosing all but its terminal 2-5 μm inside a second, insulating micropipette (Armstrong et al. 1977).

Because of the high resistance of ion-sensitive microelectrodes (10^9-10^{11} Ω), the electrode EMF must be measured with an electrometer having very high input impedance ($> 10^{13}$ Ω). The Model FD 223 electrometer (WP Instruments, New Haven, Conn.) is particularly well suited for ion-sensitive microelectrode work. All equipment, including the Faraday cage, is grounded to the steel plate (Fig. 3.6b). A low-pass filter consisting of a 15-KΩ resistor and 2-mF capacitor is placed at the electrometer output to remove high-frequency noise. The filtered signal is recorded on a strip chart recorder or may be tape-recorded or digitized and stored in computer memory. The positioning of the electrodes is shown in Fig. 3.7a.

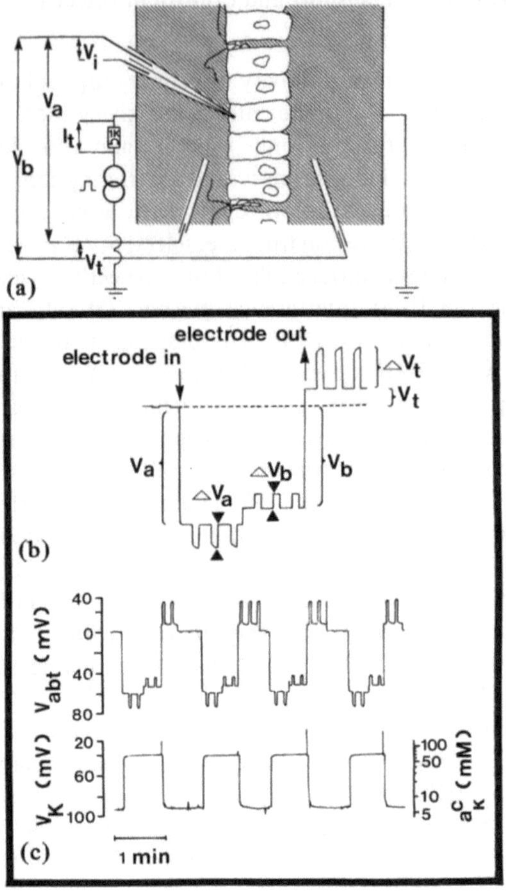

Fig. 3.7. Measurement of intracellular ion activity by double-barreled, ion-selective electrode. **(a)** Agar bridge electrodes are positioned on both sides of a rectal epithelium and a double-barreled, ion-sensitive microelectrode is located intracellularly, with the tissue mounted as shown in Fig. 3.6a. The difference between the intracellular reference and ion-sensitive barrels (V_i) indicates intracellular activity of the ion in question. The difference between the intracellular reference barrel and the two external electrodes gives the potential difference across the apical (V_a, lumen-facing) and basal (V_b, hemocoel-facing) plasma membranes, respectively. The difference between the two external electrodes indicates transrectal potential (V_t). **(b)** Recording sequence used when locust rectal cells are impaled from the lumen side (cuticle removed) and bathed in complex saline B (Table 3.2). Square-wave current pulses (I_t) are applied across the rectal wall (described in Fig. 3.6) and the resulting voltage deflections across apical and basal membranes (ΔV_a, ΔV_b) and the whole rectal wall (ΔV_t) are used to calculate membrane conductances and voltage-divider ratios. **(c)** Upper trace: actual recording of V_a, V_b, V_t, ΔV_a, ΔV_b, and ΔV_t according to the sequence shown in part b. Lower trace: the corresponding recording of V_i with values of intracellular K activity expressed as the differential intracellular electrode potential ($V_K = V_i$) and the corresponding K on ion activity (a^c_K) in millimolars. [From Hanrahan JW and Phillips JE (1982) Mechanism and control of salt absorption in locust rectum. Am J Physiol **244**:R131–R142]

The parameters we have monitored are shown in Fig. 3.7b (Hanrahan 1982). In addition to the ion-sensitive trace (V_i), both apical and basal membrane potentials (V_a and V_b, respectively, recorded from the reference barrel of a double-barreled microelectrode) and either transepithelial potential (V_t) or I_{sc} are also recorded. Voltage divider ratios (α; the ratio of apical-to-basal membrane resistance) and transepithelial resistance (R_t) are routinely calculated during impalements from the deflections in V_a, V_b, and V_t produced by transepithelial current pulses (20 μA, approximately 0.5 Hz, 1-sec duration), which are passed between the silver-foil electrodes at both ends of the chamber (Fig. 3.6a). Both α and R_t must be corrected for series resistance by subtracting voltage deflections that are observed without any tissue present.

Typical recordings of apical membrane potential (V_a) and intracellular ion activity (V_K) measured with double-barreled K-sensitive microelectrodes are shown in Fig. 3.7c. Intracellular K measurements are then used to correct for the imperfect selectivity of the Na microelectrode. Successful impalements are characterized by an abrupt, monotonic deflection in electrode EMF, a stable intracellular potential (±1 mV), a constant voltage-divider ratio during the impalement, and a return to original baseline potential upon retraction of the microelectrode. Calibration curves were usually identical before and after a series of 12 impalements.

The net electrochemical gradient of an ion i (in millivolts) across the apical or basal cell border is

$$\Delta\bar{\mu}_i/F = RT(\ln a_i^{c} - \ln a_i^{m,s})F + ZV_{a,b} \qquad (3.4)$$

where a_i^c is the intracellular activity of ion i, $a_i^{m,s}$ is the activity of ion i in the mucosal or serosal solution, F is the Faraday constant (96486.8 coulombs · mole^{-1}, R is the gas constant (8.314 joule · degree^{-1} · mole^{-1}), T is the absolute temperature in degrees Kelvin and Z is the charge on ion i. If a particular electrochemical gradient is in the opposite direction to net flux as determined by using tracers, then an active transport process at that membrane is indicated. Alternatively, $\Delta\bar{\mu}_i/F$ may favor passive diffusion in the direction of net flux. In this case, if the ion moves by electrodiffusion, then ion-sensitive microelectrodes and tracers may be used to estimate membrane conductance of the ion. For example, if net K entry occurs through the apical membrane passively by electrodiffusion, then the partial ionic conductance (G_a^K) is given by (see Hodgkin and Horowicz 1959)

$$G_a^K = J_{m\to c}^K F \left(\frac{RT}{F}\ln\frac{a_K^m}{a_K^c} - V_a\right)^{-1} \qquad (3.5)$$

where $J_{m\to c}^K$ is the influx cell from mucosa to cellular compartments measured by tracers under steady-state conditions, a_K^m and a_K^c are the activity of K in mucosal and cellular compartments, respectively, V_a is the apical membrane potential with respect to the mucosal side. R, T, and F have the same meanings as described above. Note that under steady-state I_{sc} conditions, *net* transepithelial flux is identical to *net* flux across both apical and basal membranes.

B. Demonstration of Cell–Cell Coupling Using Dyes

Intracellular injection of fluorescent dye has become a standard technique for determining whether epithelial cells are interconnected by low-resistance pathways (Loewenstein et al. 1965; reviewed by Loewenstein 1981). Lucifer Yellow CH is a relatively nontoxic, naphthalimide dye that fluoresces intensely, is small enough to pass through intercellular connections (molecular weight 457.3 daltons), and can be injected by passing current through a microelectrode into an impaled cell (Stewart 1978).

A column of dye 3-5 mm long is placed into the tip of a conventional single-barreled microelectrode (Miller 1979; Strausfeld and Miller 1980) by immersing the blunt end of the microelectrode into a 5% solution of Lucifer Yellow CH in distilled water (0.1 *M*), and then backfilled with 0.5 *M* lithium chloride. Microelectrode resistance is typically 25-40 MΩ before, and 10-15 MΩ after tips have been beveled in a jet stream of abrasive (Sect. IV.3.A.a).

To pass hyperpolarizing current (I_0) through the microelectrode, we use a constant-current amplifier (M701, WP Instruments, New Haven, Conn.) driven by wave form and pulse generators (Type 160 series, Tektronix, Beaverton, Ore.; see also Miller 1979 pp. 47-56). It is advisable to use a second electrometer to measure the amount of current actually flowing through the microelectrode because the microelectrode tip occasionally becomes blocked by dye particles, causing I_0 to decline to near zero. This problem is often remedied by switching from direct current to pulse mode (1-sec duration, 0.5 Hz) for 0.5-1.0 min. Insect epithelial cells can withstand large hyperpolarizing currents: Loewenstein et al. (1965) injected 100 nA into *Chironomus* salivary gland cells (fourth instar and prepupal) and 50 nA into *Chironomus* Malpighian tubule cells. No adverse effects were observed when 300-nA pulses of depolarizing current were injected into locust rectal cells (Hanrahan 1982). During dye injection, tissues are perfused on both sides with oxygenated saline and injected with −5 to −50 nA for 30-45 min (Hanrahan 1982).

For processing, recta are pinned onto thin wafers (2.3 cm^2) of polymerized Sylgard 184 (Dow Corning, Midland, Mich.) that have a hole cut in the center (0.2 cm^2). The purpose of pinning cut tissues is to ensure that they remain flat while being fixed in 4% paraformaldehyde buffered with 0.1 *M* phosphate at pH 7.2. After at least 1 h of fixation, tissues are dehydrated sequentially in 50%, 80%, and 100% ethanol for 5, 5, and 10 min, respectively. After clearing for 5 min in methyl salicylate, fluorescent cells are observed in whole mounts with incident light excitation at 125X or 500X magnification (Orthoplan, Leitz, Wetzler, F.R.G.; filter block I2, 50-W mercury bulb). Photomicrographs are taken using an automatic camera attachment (Orthomat W, Leitz) and 35-mm Kodak Ektachrome 160 (tungsten) Film.

If dye moves between cells, then they are almost certainly electrically coupled. However, there are instances in which cells are connected electrically but do not show dye coupling (reviewed by Loewenstein, 1981). Consequently, negative results with dye injection should be confirmed by direct electrical measurements.

C. Electrical Coupling and Flat-Sheet Cable Analysis

To measure electrical coupling between cells, current pulses (200 nA, 0.3-Hz frequency, 1-sec duration) are passed intracellularly through one single-barreled microelectrode, and the resulting voltage deflections in neighboring cells are measured by a second microelectrode (Fig. 3.8a). The distance between current-injecting and voltage-sensing microelectrodes is measured with a calibrated eye-piece micrometer. Voltage responses are displayed on a storage oscilloscope after filtering (3 dB at 5Hz) and are also recorded on an oscillographic pen recorder (7402A, Hewlett-Packard, San Diego, Calif.). In order to change the membrane potential of adjacent cells, injected current must flow between them; i.e., cells must be connected by low-resistance pathways.

It is necessary to measure the passive permeability properties of the epithelial sheet in order to fully understand transepithelial transport. Epithelia are classified as "leaky" or "tight" according to the relative resistances of the transcellular and paracellular routes to ion diffusion. Leaky epithelia are those in which most passive ion movements occur through so-called tight junctions and lateral intercellular spaces rather than across epithelial cell membranes. Leaky epithelia are usually characterized by having (1) high rates of active transport, (2) smaller transepithelial chemical and electrical potential differences as compared to tight epithelia, (3) isosmotic fluid transport, and (4) lower electrical resistance (5–300 Ω cm^2) than tight epithelia (1000–70 000 Ωcm^2) (see summary table in Frömter and Diamond 1972). Insect tissues have characteristics that are typical of both leaky and tight epithelia: high transport rates, large transepithelial gradients, and relatively low electrical resistance are found in midgut, salivary glands, and rectum.

If cells are normally uncoupled (e.g., midguts from leaf-fed *Hyalophora cecropia* or *Manduca sexta*; Blankemeyer and Harvey 1978) or if cells may be temporarily uncoupled by the injection of depolarizing current (Lewis et al. 1976), then input resistance of an individual cell (R_c) may be calculated from the deflection in apical membrane potential (ΔV_a) produced by intracellular current injection (I_0):

$$R_c = \Delta V_a / I_0 \tag{3.6}$$

To obtain the "effective" input resistance of the epithelium (R_z) in ohms square centimeters, R_c is multiplied by the total membrane area of one cell (from electron micrographs). Apical and basal membrane and junctional resistances (R_a, R_b, and R_j, respectively) are then calculated from the simple "lumped" circuit model shown in Fig. 3.8b:

$$R_a = R_z(\alpha + 1) \tag{3.7}$$

$$R_b = R_z(\alpha + 1)/\alpha \tag{3.8}$$

$$R_j = (R_t R_a + R_t R_b)/(R_a + R_b - R_t) \tag{3.9}$$

Lewis et al. (1976) found that resistances calculated by this method agreed well with those obtained by several other methods.

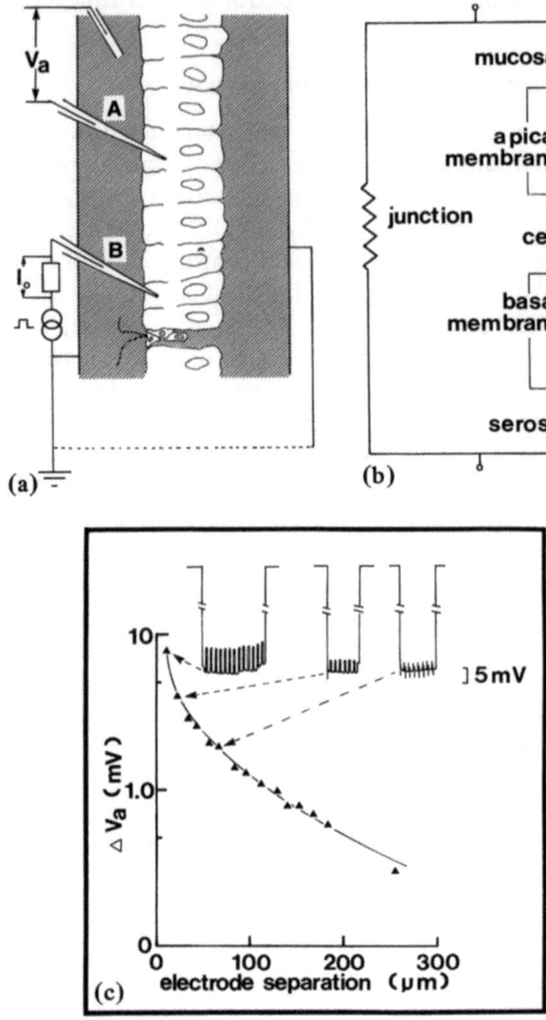

Fig. 3.8. Method used to estimate membrane resistances by flat-sheet cable analysis. **(a)** Positions of extracellular and two intracellular electrodes (A and B) during flat-sheet cable analysis of locust rectum mounted as shown in Fig. 3.6. **(b)** Circuit diagram. **(c)** Spread of current. Current pulses introduced into one cell through intracellular electrode B are recorded from intracellular electrode A located in a nearby cell. This is repeated as electrode A is moved increasing distances from electrode B to obtain the results shown graphically in part c. Detection of voltage deflections by electrode A indicates electrical coupling between rectal cells: the radial spread of current shown in part c is described by the space constant, as defined in the text. These data are used to characterize the rectal epithelium assuming the equivalent electrical circuit for rectal pad cells shown in part b. This analysis also permits evaluation of the relative importance of ion movements by transcellular versus paracellular routes. The results indicate that locust rectum is a tight epithelium of low transcellular resistance [From Hanrahan JW (1982) Cellular mechanism and regulation of KCl transport across an insect epithelium. Doctoral dissertation, University of British Columbia, Vancouver, Canada]

For a detailed description of the electrical properties of microelectrodes, see Geddes (1972). Note that mucosal and serosal sides are effectively short-circuited during intracellular injection of current because the total resistance of the epithelium is negligible compared to the basal membrane resistance of an uncoupled cell, and is still very low in the region of current flow even if cells are coupled (< 0.5% in locust rectum; see also Frömter 1972; Lewis et al. 1976).

If epithelial cells are strongly coupled, the above method cannot be used; however, junctional resistances may be determined by flat-sheet cable analysis (Eisenberg and Johnson 1970; Frömter 1972; Lewis et al. 1976; Reuss and Finn 1974, 1975; Shiba 1971; Spenney et al. 1974). In this approach, the epithelial sheet is considered as a thin, flat conductor of infinite dimensions. The radial spread of current from the site of injection may be described by the equation

$$\frac{d^2V}{dx^2} + \frac{1}{x}\frac{dV}{dx} - \frac{V}{\lambda^2} = 0 \qquad (3.10)$$

where V is the voltage deflection at some distance x, and λ is the space constant in micrometers, defined as $(R_z/R_x)^{\frac{1}{2}}$. R_z is the effective input resistance (or resistance to ground from the cell) in ohms square centimeters, and R_x is the resistance to current flow within the epithelium in kiloohms. Under the condition $V \rightarrow 0$ at $x = \infty$, the solution of Eq. (3.10) is

$$V = AK_0(x/\lambda) \qquad (3.11)$$

where K_0 is the zero-order modified Bessel function (Olver, 1967), A is an integration constant (millivolts), and x is the distance that separates current-injecting and voltage-sensing microelectrodes.

To measure the intraepithelial spread of current, voltage deflections are measured (with a storage oscilloscope) as a function of distance away from the point of intracellular current injection. Electrode separation is measured with a calibrated eyepiece micrometer. Data are then plotted semi-logarithmically and fitted to a series of curves calculated from published tables of K_0 for different values of λ (Olver 1967; see Frömter 1972) or, alternatively, by least squares curve fitting (Lewis et al. 1976). Since experimentally observed values of λ ranged between 150 and 600 μm in locust recta under various conditions, it is necessary to plot K_0 over an even wider range of λ (50–800 μm). The value of A is the vertical displacement required to fit the data points to the theoretical curve. Once A and λ are known, R_z is calculated as

$$R_z = 2\pi A\lambda^2/I_0 \qquad (3.12)$$

where I_0 is the current injected intracellularly in microamperes, and other symbols are as described above. Equations (3.7)–(3.9) are then solved using α and R_t, which are measured as in Sect. IV.3.A.B.

In Fig. 3.8c the deflections in V_a measured in locust rectum are shown as a function of distance from the point of intracellular current injection during exposure to normal saline with 1 mM cAMP. The solid line indicates the best fitting Bessel function ($\lambda = 175$ μm, $A = 1.6$ mV). These data were then combined with α and R_t to calculate individual membrane and tight junctional resistances. Locust rectum is a tight epithelium with relatively low electrical resistance, particularly during cAMP stimulation, when 95% of the transepithelial conductance is located in the transcellular pathway (Hanrahan et al. 1982).

D. Advantages and Limitations

Intracellular double-barreled microelectrodes have been used to measure Cl, K, and Na activities in locust rectum (Hanrahan 1982; Hanrahan and Phillips 1980b, 1982), K and Cl in blowfly salivary glands (Berridge 1980; Berridge and Schlue 1978), and K in lepidopteran midgut (Blankemeyer and Duncan 1980; Moffett et al. 1982; Moffett 1979). Some investigators prefer single-barreled microelectrodes because they are less likely to cause impalement damage in vertebrate epithelia (Delong and Civan 1978; Garcia-Diaz and Armstrong 1980; Reuss and Weinman 1979; see Brown and Flaming 1977, for techniques in preparing conventional microelectrodes for use in small cells). This problem is not serious in insect cells because of their large volume (the volume of a locust rectal cell is about five times that of a *Necturus* gallbladder cell) and also because of extensive membrane infolding. In locust rectum, microscopic infolding results in a 9- to 200-fold amplification of membrane area in different regions of the cell according to electron micrographs.

In the absence of impalement artifacts, double-barreled microelectrodes are superior because they measure membrane potential and intracellular ion activity simultaneously in the same cell, whereas average membrane potential must be used to calculate ion activities with single-barreled microelectrodes. Single-barreled microelectrodes cannot be used in epithelia such as *Manduca* midgut, where cells are not coupled and there is more than one population of cells with different potentials (Blankemeyer and Harvey 1978). Moreover, accurate measurements using single-barreled microelectrodes are difficult if membrane potentials are not constant (i.e., during time course experiments), and under I_{sc} conditions, when a variable voltage drop due to external series resistance contributes to measured potentials.

Sources of error in ion-sensitive microelectrode studies have been described by Durst (1974) and Armstrong and Garcia-Diaz (1980). One problem is that the electrochemical potential between the cytoplasm and bathing saline is not necessarily a transmembrane gradient *per se,* due to the complexity of the cell border. High salt concentrations have been observed in the lateral spaces of cockroach rectum by means of micropuncture (Wall and Oschman 1970) and electron probe microanalysis (Gupta et al. 1980). Although local salt gradients should not affect conclusions regarding the overall active or passive nature of transport between cytoplasm and the serosal side, the approach described here provides no insight into the question of localized ion recycling as proposed for the basolateral border of insect papillate recta (Phillips 1970; Wall and Oschman 1970). One approach might be to measure ion activities in the lateral spaces at different rates of fluid transport using ion-sensitive microelectrodes. Curci and Frömter (1979) measured K activity in the lateral intercellular spaces of *Necturus* gallbladder. The microelectrode design used in their study featured a dye-ejecting barrel for determining the exact location of the microelectrode tip (for details of construction, see Frömter et al. 1981). Studies by Wall and Oschman (1970) indicated that this is even more feasible with papillate recta, which have large intercellular sinuses that can be visualized by dyes absorbed from the lumen.

The spread of dye from an injected cell into adjacent cells demonstrates convincingly that they are functionally connected by low-resistance pathways. Loewenstein (1981) listed 47 different molecules that have been injected into cells to test cell-cell coupling. Lucifer Yellow CH is extremely fluorescent (700 times more intense than procion yellow) and fades negligibly in injected cells if stored in the dark and refrigerated. This dye is readily available (Sigma Chemical Co., St. Louis, Mo.), does not penetrate normal cell membranes (in contrast to fluorescein), spreads within seconds, is relatively nontoxic, and is easily fixed with a variety of different fixatives (Stewart 1978).

Cell uncoupling may result from unnaturally high CO_2 tension *in vitro* (Turin and Warner 1977) or may be an artifact of depolarizing (Socolar and Politoff 1971) or damaging the epithelial cells. These possibilities must be considered when coupling is not observed. For example, the trauma associated with dissecting the cuticular intima off the mucosal surface of locust rectum may uncouple epithelial cells at the edge of the rectal pads (Hanrahan 1982).

In order to apply flat-sheet cable analysis, it is necessary to use a very simple circuit model to describe the electrical properties of epithelia having complex ultrastructure, particularly in insects. It must be assumed that there is no EMF between the lateral intercellular space and the external bathing media. Salt gradients between the lateral spaces and bathing saline generate an EMF across the epithelium if the selective permeabilities of junctions and lateral spaces are different. As mentioned above, there is considerable evidence that intercellular spaces in insect epithelia have high salt concentrations, at least in freshly dissected tissues. However, no electrophysiological evidence for an EMF in the lateral space was obtained in locust rectum perfused *in vitro* (Hanrahan 1982). The relationship between function and ultrastructure is not clear in insect rectum: rectal cells are held together laterally by numerous junctions from the mucosal to serosal ends of the intercellular spaces (Lane 1979; Wall and Oschman 1975). As a result, there is uncertainty as to how far the apical and basal cell membranes extend into the lateral intercellular space and also whether the lateral membrane is functionally part of apical or basal membranes.

The simple circuit model shown in Fig. 3.8b ignores the effects of current that flows through the lateral intercellular space. If this pathway is significant in a particular epithelium, then a "distributed" circuit model (Boulpaep and Sackin 1980; Clausen et al. 1979) is more accurate than the "lumped" model but it requires more circuit parameters than can be obtained by cable analysis alone (see Kottra and Frömter 1983).

V. Conclusion

A variety of hindgut preparations have been useful in studying transport processes in the locust rectum. Generally, *in situ* preparations are suitable for identifying active transport processes and for studying their overall regulation and physiological significance, whereas *in vitro* techniques are required for more detailed studies of cellular mechanisms. Extending these studies to other insect species will contribute greatly to our understanding of hindgut function.

Acknowledgments. This work was supported by operating grants from Natural Sciences and Engineering Research Council of Canada to J. Phillips, and an NSERC Post-Graduate Scholarship to J. Hanrahan.

References

Allen RJL (1940) The estimation of phosphorus. Biochem J **34**:858–865

Andrusiak EW (1974) Resorption of phosphate, calcium, and magnesium in the *in vitro* locust rectum. Master's thesis, University of British Columbia, Vancouver, Canada

Andrusiak EW, Phillips JE, Speight J (1980) Phosphate transport by locust rectum *in vitro*. Can J Zool **58**:1518–1523

Anstee, JH, Bell DM, Fathpour H (1979) Fluid and cation secretion by the Malpighian tubules of *Locusta*. J Insect Physiol **25**:373–380

Armstrong WMcD, Garcia-Diaz JF (1980) Ion selective microelectrodes: Theory and technique. Fed Proc **39**:2851–2859

Armstrong WM, Wojtkowski W, Bixenman WR (1977) A new solid-state microelectrode for measuring intracellular chloride activities. Biochim Biophys Acta **465**:165–170

Balshin M, Phillips JE (1971) Active absorption of amino-acids in the rectum of the desert locust (*Schistocerca gregaria*). Nature (New Biol) **233**:53–55

Bates RG (1973) Determination of pH; Theory and practice, 2nd ed. Wiley, New York

Baumeister T, Meredith J, Julien W, Phillips J (1981) Acetate transport by locust rectum *in vivo*. J Insect Physiol **27**:195–201

Baumgarten CM (1981) An improved liquid ion exchanger for chloride ion-selective microelectrodes. Am J Physiol **241**:C258–C263

Berridge MJ (1966) Metabolic pathways of isolated Malpighian tubules of the blowfly functioning in an artificial medium. J Insect Physiol **12**:1523–1538

Berridge MJ (1980) The role of cyclic nucleotides and calcium in the regulation of chloride transport. Ann NY Acad Sci **341**:156–169

Berridge MJ, Schlue WR (1978) Ion-selective electrode studies on the effects of 5-hydroxytryptamine on the intracellular level of potassium in an insect salivary gland. J Exp Biol **72**:203–216

Bindslev N, Hansen AJ (1981) Mono-/bivalent ion selectivities obtained by the Nicolsky and the electrodiffusional regimes. In: Lübbers DW, Acker H,

Buck RP, Eisenman G, Kessler M, Simon W (eds) Progress in enzyme and ion-selective electrodes. Springer, New York, pp 25–31

Blankemeyer JT (1977) The route of active potassium ion transport in the midgut of *Hyalophora cecropia* and *Manduca sexta.* Doctoral dissertation, Temple University, Philadelphia

Blankemeyer JT (1978) Demonstration of a pump-mediated efflux in the epithelial potassium active transport system of insect midgut. Biophys J **23**:313–318

Blankemeyer JT, Duncan RL (1980) The potassium activity in a polymorphic potassium active transporting epithelium, insect midgut. Fed Proc **39**:1711

Blankemeyer JT, Harvey WR (1978) Identification of active cell in potassium transporting epithelium. J Exp Biol **77**:1–13

Bolton TB, Vaughan-Jones RD (1977) Continuous direct measurement of intracellular chloride and pH in frog skeletal muscle. J Physiol (Lond) **270**:801–833

Bonventre JV, Blouch K, Lechene C (1980) Liquid droplets and isolated cells. In: Hayat HA (ed) X-ray microanalysis in biology. University Park Press, Baltimore, pp 307–366

Boulpaep EL, Giebisch G (1978) Electrophysiological measurements on the renal tubule. In: Martinez-Maldonado M (ed) Methods in pharmacology, Vol 4B. Plenum Press, New York, pp 165–193

Boulpaep EL, Sackin H (1980) Electrical analysis of intraepithelial barriers. In: Bronner F, Kleinzeller A (eds) Current topics in membranes and transport, Vol 13. Academic Press, New York, pp 169–197

Brodsky AB (1978) CRC handbook of radiation measurement and protection, Sect A, Vol I: Physical science and engineering data. CRC Press, West Palm Beach, Fla, pp 340–356

Brown HM (1976) Intracellular Na^+, K^+ and Cl^- activities in *Balanus* photoreceptors. J Gen Physiol **68**:281–296

Brown KT, Flaming DG (1977) New microelectrode techniques for intracellular work in small cells. Neuroscience **2**:813–827

Buck JB (1953) Physical properties and chemical composition of insect blood. In: Roeder KD (ed) Insect physiology. Wiley, New York, pp 147–190

Burns DT, Coy JS, Hayes WP, Kent DM (1974) The indirect spectophotometric determination of sulphate with 2-aminoperimidine hydrochloride. Mikrochim Acta [Wien] 245–248

Burton, RF (1975) Ringer solutions and physiological salines. Wright, Bristol, England

Caflisch CR, Carter NW (1974) A micro P_{CO_2} electrode. Anal Biochem **60**:252–257

Caflisch CR, Pucacco LR, Carter NW (1978) Manufacture and utilization of antimony pH electrodes. Kidney Int **14**:126–141

Chamberlin M (1981) Metabolic studies on the locust rectum. Doctoral dissertation, University of British Columbia, Vancouver, Canada

Chamberlin ME, Phillips JE (1980) Proline transport by locust Malpighian tubules. Am Zool **20**:945

Chamberlin ME, Phillips JE (1982) Regulation of hemolymph amino acid levels and active secretion of proline by Malpighian tubules of locusts. Can J Zool **60**:2745–2752

Chonko AM, Irish JM III, Welling DJ (1978) Microperfusion of isolated tubules. In: Martinez-Maldonado M (ed) Methods in pharmacology, Vol 4B. Plenum Press, New York, pp 221–258

Clausen C, Lewis SA, Diamond JM (1979) Impedance analysis of a tight epithelium using a distributed resistance model. Biophys J **26**:291–318

Clements AN, May TE (1974) Studies on locust neuromuscular physiology in relation to glutamic acid. J Exp Biol **60**:673–705

Cooper PD, Deaton LE, Jungreis AM (1980) Chloride transport during resorption of molting fluid across the pharate pupal integument of tobacco hornworms, *Manduca sexta*. J Gen Physiol **76**:13a–14a

Cornell JC, Jungreis AM (1981) Changes in K transport across the isolated integument of the tobacco hornworm *Manduca sexta*. Am Zool **21**:997

Curci S, Frömter E (1979) Micropuncture of lateral intercellular spaces of *Necturus* gallbladder to determine space fluid K^+ concentration. Nature **278**:355–357

Dantzler WH (1977) *In vitro* microperfusion. In: Gupta BL, Moreton RB, Oschman JL, Wall BJ (eds) Transport of ions and water in animals. Academic Press, New York, pp 57–82

Delong J, Civan MM (1978) Dissociation of cellular K^+ accumulation from net Na^+ transport by toad urinary bladder. J Membr Biol **42**:19–43

DeSousa RC, Li JH, Essig A (1971) Flux ratios and isotope interaction in an ion exchange membrane. Nature **231**:44–45

Dow JAT (1981a) Ion and water transport in locust alimentary canal: Evidence from *in vivo* electrochemical gradients. J Exp Biol **93**:167–179

Dow JAT (1981b) Localization and characterization of water uptake from the midgut of the locust, *Schistocerca gregaria*. J Exp Biol **93**:269–281

Duchâteau G, Florkin M, Leclercq J (1953) Concentrations des bases fixes et types de composition de la base totale de l'hémolymphe des insectes. Arch Int Physiol **61**:518–549

Duffey ME, Turnheim K, Frizzell RA, Schultz SG (1978) Intracellular chloride activities in rabbit gallbladder: Direct evidence for the role of the sodium-gradient in energizing "uphill" chloride transport. J Membr Biol **42**:229–245

Durst RA (1974) Ion-selective electrode response in biologic fluids. In: Berman HJ, Hebert NC (eds) Ion-selective microelectrodes. Plenum Press, New York, pp 13–21

Ehrlich BE, Diamond JM (1978) An ultramicro method for analysis of lithium and other biologically important cations. Biochem Biophys Acta **543**:264–268

Eisenberg RS, Johnson EA (1970) Three-dimensional electrical field problems in physiology. In: Butler JAV, Noble D (eds) Progress in biophysics and molecular biology, Vol 20. Pergamon Press, Toronto, pp 1–65

Ernster L, Zetterström R, Lindberg O (1950) A method for the determination of tracer phosphate in biological material. Acta Chem Scand **4**:942–947

Farmer J, Maddrell SHP, Spring JH (1981) Absorption of fluid by the midgut of *Rhodnius*. J Exp Biol **94**:301–316

Flemström G, Öberg PA, Petterson H (1973) A new device for automatic measurement of short-circuit current across epithelial tissues. Ups J Med Sci **78**:19–21

Fletcher CR (1978) Improved flame photometry. J Exp Biol **77**:243–246

Florkin M, Jeuniaux C (1974) Hemolymph: Composition. In: Rockstein M (ed) The physiology of insecta, 2nd ed. Vol V. Academic Press, New York, pp 255–307

Fromm M, Schultz SG (1981) Some properties of KCl-filled microelectrodes: Correlation of potassium "leakage" with tip resistance. J Membr Biol **62**:239–244

Frömter E (1972) The route of passive ion movement through the epithelium of *Necturus* gallbladder. J Membr Biol **8**:259–301

Frömter E, Diamond J (1972) Route of passive ion permeation in epithelia. Nature (New Biol) **235**:9–13

Frömter E, Simon M, Gebler B (1981) A double-channel ion-selective microelectrode with the possibility of fluid ejection for localization of the electrode tip in the tissue. In: Lübbers DW, Acker B, Buck RP, Eisenman G, Kessler M, Simon W (eds) Progress in enzyme and ion-selective electrodes. Springer, New York, pp 35–44

Fujimoto M, Kubota T (1976) Physicochemical properties of a liquid ion exchanger microelectrode and its application to biological fluid. Japn J Physiol **26**:631–650

Fujimoto M, Naito K, Kubota T (1980) Electrochemical profile for ion transport across the membrane of proximal tubular cells. Membr Biochem **3**:67–97

Garcia-Diaz JF, Armstrong WMcD (1980) The steady-state relationship between sodium and chloride transmembrane electrochemical potential differences in *Necturus* gallbladder. J Membr Biol **55**:213–222

Garland HO, Brown JA, Henderson IW (1978) X-ray analysis applied to the study of renal tubular fluid samples. In: Erasmus DA (ed) Electron probe microanalysis in biology. Chapman and Hall, London

Geddes LA (1972) Electrodes and the measurement of bioelectric events. Wiley, New York

Giebisch G (ed) (1972) Renal micropuncture techniques: A symposium. Yale J Biol Med **45**:187–456

Giebisch G (1977) Micropuncture techniques. In: Gupta BL, Moreton RB, Oschman JL, Wall BJ (eds) Transport of ions and water. Academic Press, New York, pp 29–56

Goh SL (1971) Mechanism of water and salt absorption in the *in vitro* locust rectum. Master's thesis, University of British Columbia, Vancouver, Canada

Goh S, Phillips JE (1978) Dependence of prolonged water absorption by *in vitro* locust rectum on ion transport. J Exp Biol **72**:25–41

Gottschalk CW, Lassiter WE (1973) Micropuncture methodology. In: Orloff J, Berliner RW (eds) Handbook of physiology, Sect 8: Renal physiology. American Physiological Society, Washington, DC, pp 129–143

Green R, Giebisch G (1974) Some problems with antimony microelectrodes. In: Berman HJ, Hebert NC (eds) Ion-selective microelectrodes. Plenum Press, New York, pp 43–53

Greger R, Lang F, Knox FG, Lechene C (1978) Analysis of tubule fluid. In: Martinez-Maldonado M (ed) Methods in pharmacology, Vol 4B. Plenum Press, New York, pp 105–140

Gupta BL, Hall TA (1979) Quantitative electron probe x-ray microanalysis of electrolyte elements within epithelial tissue compartments. Fed Proc **38**: 144–153

Gupta BL, Hall TA (1981) The x-ray microanalysis of frozen-hydrated sections in scanning electron microscopy: An evaluation. Tissue Cell **13**:623–643

Gupta BL, Hall TA, Moreton RB (1977) Electron probe x-ray microanalysis. In: Gupta BL, Moreton RB, Oschman JL, Wall BJ (eds) Transport of ions and water in animals. Academic Press, New York, pp 83–168

Gupta BL, Wall BJ, Oschman JL, Hall TA (1980) Direct microprobe evidence of local concentration gradients and recycling of electrolytes during fluid absorption in the rectal papillae of *Calliphora.* J Exp Biol **88**:21–47

Hanrahan JW (1978) Hormonal regulation of chloride in locusts. Physiologist **21**:50

Hanrahan JW (1982) Cellular mechanism and regulation of KCl transport across an insect epithelium. Doctoral dissertation, University of British Columbia, Vancouver, Canada

Hanrahan JW, Phillips JE (1980a) Na^+-independent Cl^- transport in an insect. Fed Proc **39**:285

Hanrahan JW, Phillips JE (1980b) Characterization of locust Cl^- transport. Am Zool **20**:938

Hanrahan JW, Phillips JE (1982) Mechanism and control of salt absorption in locust rectum. Am J Physiol **244**:R131–R142

Hanrahan JW, Phillips JE, Steeves JD (1982) Electrophysiology of Cl transport across insect rectum: Effects of cAMP. Fed Proc **41**:1496

Harvey WR, Nedergaard S (1964) Sodium-independent active transport of potassium in the isolated midgut of the Cecrepia silkworm. Proc Natl Acad Sci USA **51**:757–765

Helman SI, Miller DA (1973) Edge damage effect on electrical measurements of frog skin. Am J Physiol **225**:972–977

Helman SI, Miller DA (1974) Edge damage effect on measurements of urea and sodium flux in frog skin. Am J Physiol **226**:1198–1203

Herrera L, Jordana R, Ponz F (1976) Chloride-dependent transmural potential in the rectal wall of *Schistocerca gregaria.* J Insect Physiol **22**:291–297

Herrera L, Jordana R, Ponz F (1977) Effect of inhibitors on chloride-dependent transmural potential in the rectal wall of *Schistocerca gregaria.* J Insect Physiol **23**:677–682

Hodge C (1939) The anatomy and histology of the alimentary tract of *Locusta migratoria* L. (Orthoptera: Acrididae). J Morphol **64**:375–400

Hodgkin AL, Horowicz P (1959) The influence of potassium and chloride ions on the membrane potential of single muscle fibres. J Physiol (Lond) **148**: 127–160

Holder RED, Sattelle DB (1972) A multiway non-return value for use in physiological experiments. J Physiol (Lond) **226**:2P–3P

Isaacson LC, Douglas RJ, Pepler J (1971) Automatic measurement of voltage and short-circuit current across amphibian epithelia. J Appl Physiol **31**:298–299

Jacquez JA (1980) Tracers in the study of membrane processes. In: Andreoli TE, Hoffman JF, Fanestil DD (eds) Membrane physiology. Plenum Press, New York, pp 147–164

Jungries AM, Harvey WR (1975) Role of active potassium transport by integumentary epithelium in secretion of larval–pupal moulting fluid during silk moth development. J Exp Biol **62**:357–366

Karlmark B (1973) The determination of titratable acid and ammonium ions in picomole amounts. Anal Biochem **52**:69–82

Karlmark B, Sohtell M (1973) The determination of bicarbonate in nanoliter samples. Anal Biochem **53**:1–11

Karlmark B, Sohtell M, Ulfendahl HR (1971) A pH-glass electrode for nanolitre biological samples. Acta Soc Med Ups **76**:58–62

Karlmark B, Jaeger P, Fein H, Giebisch G (1982) Coulometric acid–base titration in nanoliter samples with glass and antimony electrodes. Am J Physiol **242**: F49–F99

Khuri RN, Agulian SK (1981) Intracellular electro-chemical studies of single renal tubule cells and muscle fibers. In: Lübbers DW, Acker B, Buck RP, Eisenman G, Kessler M, Simon W (eds) Progress in enzyme and ion-selective electrodes. Springer, New York, pp 195–205

Khuri RN, Agulian SK, Oelert H, Harik RI (1967) A single unit pH glass ultramicro electrode. Pfluegers Arch **294**:291–294

Khuri RN, Agulian SK, Kalloghlian A (1972) Intracellular potassium in cells of the distal tubule. Pfluegers Arch **335**:297–308

Khuri RN, Bogharian K, Agulian SK (1974) Intracellular bicarbonate in single skeletal muscle fibres. Pfluegers Arch **349**:285–294

King EJ (1932) The colorimetric determination of phosphorus. Biochem J **26**:292–297

Kotera K, Satake N, Honda M, Fujimoto M (1979) The measurement of intracellular sodium activities in the bullfrog by means of a double-barreled sodium liquid ion-exchange microelectrode. Membr Biochem **2**:232–338

Kottra G, Frömter E (1983) Functional properties of the paracellular pathway in some leaky epithelia. J Exp Biol in press

Küppers J, Thurm U (1980) Water transport by electroosmosis. In: Locke M, Smith DS (eds) Insect biology in the Future, "VBW 80." Academic Press, New York, pp. 125–144

LaForce RC (1967) Device to measure the voltage–current relations in biological membranes. Rev Sci Instrum **38**:1225–1228

Lane NJ (1979) Freeze-fracture and tracer studies on the intercellular junction of insect rectal tissues. Tissue Cell **11**:481–506

Leader JP, Green LB (1978) Active transport of chloride and sodium by the rectal chamber of the larvae of the dragonfly, *Uropetala carovei.* J Insect Physiol **24**:685–692

Lechene C, Warner RR (1979) Electron probe analysis of liquid droplets. In: Lechene C, Warner RR (eds) Microbeam analysis in biology. Academic Press, New York, pp 279–298

Lee CO, Taylor A, Windhager EE (1980) Cytosolic calcium ion activity in epithelial cells of *Necturus* kidney. Nature **287**:859–861

Lee RM (1961) The variation of blood volume with age in the desert locust (*Schistocerca gregaria*). J Insect Physiol **6**:36–51

Lettau J, Foster WA, Harker JE, Treherne JE (1977) Diel changes in potassium activity in the hemolymph of the cockroach *Leucophaea maderae.* J Exp Biol **71**:171–186

Levenbook L (1950) The composition of horse bot fly (*Gastrophilus intestinalis*) larva blood. Biochem J **47**:336–346

Levine DZ (1972) Measurement of tubular fluid bicarbonate concentration by the cuvette-type glass micro pH electrode. Yale J Biol Med **45**:368–372

Lewis SA, Diamond JM (1976) Na transport by rabbit urinary bladder, a tight epithelium. J Membr Biol **28**:1–40

Lewis SA, Wills NK (1980) Resistive artifacts in liquid-ion exchanger microelectrode estimates of Na^+ activity in epithelial cells. Biophys J **31**:127–138

Lewis SA, Eaton DC, Diamond JM (1976) The mechanism of Na^+ transport by rabbit urinary bladder. J Membr Biol **28**:41–70

Lewis SA, Wills NK, Eaton DC (1978) Basolateral membrane potential of a tight epithelium: Ionic diffusion and electrogenic pumps. J Membr Biol **41**:117–148

Li JH, DeSousa RC, Essig A (1974) Kinetics of tracer flows and isotope interaction in an ion exchange membrane. J Membr Biol **93**:104

Lindemann B (1975) Impalement artifacts in microelectrode recordings of epithelial membrane potentials. Biophys J **15**:1161–1164

Little C (1974) A method for determining total carbon dioxide in nanolitre volumes of liquid. J Exp Biol **61**:667–675

Little C (1977) Microsample analysis. In: Gupta BL, Moreton RB, Oschman JL, Wall BJ (eds) Transport of ions and water in animals. Academic Press, New York, pp 15–28

Loewenstein WR (1981) Junctional intercellular communication: The cell-to-cell membrane channel. Physiol Rev **61**:829–913

Loewenstein WR, Socolar SJ, Higashino S, Kanno Y, Davidson N (1965) Intercellular communication: Renal, urinary bladder, sensory, and salivary gland cells. Science **149**:295–298

Maddrell SHP (1971) The mechanisms of insect excretory systems. In: Beament JWL, Treherne JE, Wigglesworth VB (eds) Advances in insect physiology, Vol 8. Academic Press, New York, pp 199–331

Maddrell SHP (1977) Insect malpighian tubules. In: Gupta BL, Moreton RB, Oschman JL, Wall BJ (eds) Transport of ions and water in animals. Academic Press, New York, pp 57–82

Maddrell SHP (1980) Bioassay of diuretic hormones in *Rhodnius.* In: Miller TA (ed) Neurohormonal techniques in insects. Springer, New York, pp 81–90

Maddrell SHP, Gardiner BOC (1980) The permeability of the cuticular lining of the insect alimentary canal. J Exp Biol **85**:227–237

Maddrell SHP, Klunsuwan S (1973) Fluid secretion by *in vitro* preparations of the Malpighian tubules of the desert locust, *Schistocerca gregaria.* J Insect Physiol **19**:1369–1376

Maffly RH (1968) Conductometric method for measuring micromolar quantities of carbon dioxide. Anal Biochem **23**:252–262

Malnic G, Vieira FL (1972) Antimony microelectrode in kidney micropuncture. Yale J Biol Med **45**:356

Miller TA (ed) (1979) Neurohormonal techniques in insects. Springer, New York

Moffett DF (1979) Potassium activity of single insect midgut cells. Am Zool **19**:996

Moffett DF, Hudson RL, Moffett SB, Ridgway RL (1982) Intracellular K^+ activities and cell membrane potentials in a K^+-transporting epithelium, the midgut of tobacco hornworm (*Manduca sexta*). J Membrane Biol **70**:59–68

Moody GJ, Thomas JDR (1971) Selective ion sensitive electrodes. Merrow, Watford, England

Mordue W (1969) Hormonal control of Malpighian tube and rectal function in the desert locust, *Schistocerca gregaria.* J Insect Physiol **15**:273–285

Mordue W (1972) Hydromineral regulation in animals. Part I. Hormones and excretion in locusts. Gen Comp Endocrinol (Suppl) **3**:289–298

Moreton RB (1981) Electron-probe x-ray microanalysis: Techniques and recent application in biology. Biol Rev **56**:409–461

Neild TO, Thomas RC (1973) New design for a chloride-sensitive microelectrode. J Physiol (Lond) **231**:7P–8P

Nelson DJ, Ehrenfeld J, Lindemann B (1978) Volume changes and potential artifacts of epithelial cells of frog skin following impalement with microelectrodes filled with 3M KCl. J Membr Biol (Special Issue) **40**:91–119

O'Doherty J, Garcia-Diaz JF, Armstrong WMcD (1979) Sodium-selective liquid ion-exchanger microelectrodes for intracellular measurements. Science **203**: 1349–1351

Ogden TE, Citron MC, Pierantoni R (1978) The jet stream microbeveler: An inexpensive way to bevel ultrafine glass micropipettes. Science **201**:469–470

Olver FWJ (1967) Bessel functions of integer order. In: Abramowitz M, Stegun JA (eds) Handbook of mathematical functions. National Bureau of Standards, Washington, DC, pp 355–422

Palmer LG, Civan MM (1977) Distribution of Na^+, K^+ and Cl^- between nucleus and cytoplasm in *Chironomus* salivary gland cells. J Membr Biol **33**:41–61

Paulson S (1953) Biophysical and physiological investigations on cartilage and other mesenchymal tissues. IV. A semimicro method for conductometric determination of sulfur. Acta Chem Scand **7**:325–328

Peng CT (1977) Sample preparation in liquid scintillation counting. Amersham Corp, Arlington Heights, Ill

Peng CT, Horrocks DL, Alpen EL (eds) (1980) Liquid scintillation counting: Recent applications and developments, Vol 1: Physical aspects. Academic Press, New York

Phillips JE (1964a) Rectal absorption in the desert locust, *Schistocerca gregaria* Forskål. I. Water. J Exp Biol **41**:15–38

Phillips JE (1964b) Rectal absorption in the desert locust, *Schistocerca gregaria* Forskål. II. Sodium, potassium and chloride. J Exp Biol **41**:39–67

Phillips JE (1964c) Rectal absorption in the desert locust, *Schistocerca gregaria* Forskål. III. The nature of the excretory process. J Exp Biol **41**:67–80

Phillips JE (1970) Apparent transport of water by insect excretory systems. Am Zool **10**:413–436

Phillips JE (1980) Epithelial transport and control in recta of terrestrial insects. In: Locke M, Smith DS (eds) Insect biology in the future. Academic Press, New York, pp 145–177

Phillips J (1981) Comparative physiology of insect renal function. Am J Physiol **241**:R241–R257

Phillips JE, Beaumont C (1971) Symmetry and non-linearity of osmotic flow across rectal cuticle of the desert locust. J Exp Biol **54**:317–328

Phillips JE, Dockrill AA (1968) Molecular sieving of hydrophilic molecules by the rectal intima of the desert locust (*Schistocerca gregaria*). J Exp Biol **48**:521–532

Phillips JE, Mordue W, Meredith J, Spring J (1980) Purification and characteristics of the chloride transport stimulating factor from the locust corpora cardiaca: A new peptide. Can J Zool **58**:1851–1860

Phillips JE, Spring J, Hanrahan J, Mordue W, Meredith J (1981) Hormonal control of salt reabsorption by the excretory system of an insect: Isolation of a new protein. In: Farner DS, Lederis K (eds) Neurosecretion: Molecules, cells, systems. Plenum Press, New York, pp 373–382

Phillips JE, Meredith J, Spring J, Chamberlin M (1982) Control of ion reabsorption in locust rectum: Implications for fluid transport. J Exp Zool in press

Pichon Y (1970) Ionic content of haemolymph in the cockroach, *Periplaneta americana*. A critical analysis. J Exp Biol **53**:195–209

Prager DJ, Bowman RL, Vurek GG (1965) Constant volume, self-filling nanolitre pipette: Construction and calibration. Science **147**:606–608

Prusch RD (1974) Active ion transport in the larval hindgut of *Sarcophaga bullata* (Diptera: Sarcophagidae). J Exp Biol **61**:95–109

Prusch RD (1976) Unidirectional ion movements in the hindgut of larval *Sarcophaga bullata* (Diptera: Sarcophagidae). J Exp Biol **64**:89–100

Puschett JB, Zurbach PE (1974) Re-evaluation of microelectrode methodology for the *in vitro* determination of pH and bicarbonate concentration. Kidney Int. **6**:81–91

Quehenberger P (1977) The influence of carbon dioxide, bicarbonate and other buffers on the potential of antimony microelectrodes. Pfluegers Arch **368**: 141–147

Quinton PM (1976) Construction of pico-nanoliter self-filling volumetric pipettes. J Appl Physiol **40**:260–262

Quinton PM (1978) Techniques for microdrop analysis of fluids (sweat, saliva, urine) with an energy-dispersive x-ray spectrometer on a scanning electron microscope. Am J Physiol **234**:F255–F259

Ramsay JA (1954) Active transport of water by the Malpighian tubules of the stick insect, *Dixippus morosus* (Orthoptera, Phasmidiae). J Exp Biol **31**: 104–113

Ramsay JA, Brown RHJ, Croghan PC (1955) Electrometric titration of chloride in small volume. J Exp Biol **32**:822–829

Rehm WS (1975) Ion transport and short-circuit technique. In: Bronner F, Kleinzeller A (eds) Current topics in membrane and transport, Vol 7. Academic Press, New York, pp 217–270

Reuss L, Finn AL (1974) Passive electrical properties of toad urinary bladder epithelium: Intercellular electrical coupling and transepithelial cellular and shunt conductances. J Gen Physiol **64**:1–25

Reuss L, Finn AL (1975) Electrical properties of the cellular transepithelial pathway in *Necturus* gallbladder. I. Circuit analysis and steady-state effects of mucosal solution ionic substitutions. J Membr Biol **25**:115-139

Reuss L, Grady TP (1979) Effects of external sodium and cell membrane potential on intracellular chloride activity in gallbladder epithelium. J Membr Biol **51**:15-31

Reuss L, Weinman SA (1979) Intracellular ionic activities and transmembrane electrochemical potential differences in gallbladder epithelium. J Membr Biol **49**:345-362

Rick R, Horster M, Dorge A, Thurau K (1977) Determination of electrolytes in small biological fluid samples using energy dispersive x-ray microanalysis. Pfluegers Arch **369**:95-98

Riegel JA (1970) *In vitro* studies of fluid and ion movements due to the swelling of formed bodies. Comp Biochem Physiol **35**:843-856

Roach DK (1963) Analysis of the haemolymph of *Arion ater* L. (Gastropoda: Pulmonata). J Exp Biol **40**:613-623

Robinson RA, Stokes RH (1965) Electrolyte solutions, 3rd ed. Butterworths, London

Roinel N (1975) Electron microprobe quantitative analysis of lyophilised 10^{-10} 1 volume samples. J Microsc Bull Cell **22**:261-268

Rothe CF, Quay JF, Armstrong WM (1969) Measurement of epithelial electrical characteristics with an automatic voltage clamp device with compensation for solution resistance. IEEE Trans Biomed Eng **16**:160-169

Saunders JH, Brown HM (1977) Liquid and solid-state Cl^--sensitive microelectrodes: Characteristics and application to intracellular Cl^- activity in *Balanus* photoreceptors. J Gen Physiol **70**:507-530

Schoen HF, Candia OA (1978) An inexpensive, high output voltage, voltage clamp for epithelial membranes. Am J Physiol **235**:C69-C72

Shaw, TI (1955) Potassium movements in washed erythrocytes. J Physiol **129**: 464-475

Shiba H (1971) Heaviside's "Bessel cable" as an electric model for flat simple epithelial cells with low resistive junctional membranes. J Theor Biol **30**: 59-68

Smith PG (1978) A device for automatic measurement of short-circuit current in epithelia. Comp Biochem Physiol **61**A:221-222

Socolar SJ, Politoff AL (1971) Uncoupling cell junctions of a glandular epithelium by depolarizing current. Science **172**:492-494

Sohtell M, Karlmark B (1976) *In vivo* micropuncture P_{CO_2} measurements. Pflugers Arch **363**:179-189

Speight J (1967) Acidification of rectal fluid in the locust *Schistocerca gregaria*. Master's thesis, University of British Columbia, Vancouver, Canada

Spenney JG, Shoemaker RL, Sachs G (1974) Microelectrode studies of fundic gastric mucosa: Cellular coupling and shunt conductance. J Membr Biol **19**:105-128

Spring JH, Phillips JE (1980a) Studies on locust rectum: I Stimulants of electrogenic ion transport. J Exp Biol **86**: 211-223

Spring JH, Phillips JE (1980b) Studies on locust rectum: II Identification of specific ion transport processes regulated by corpora cardiacum and cyclic-AMP. J Exp Biol **86**:225-236

Spring JH, Phillips JE (1980c) Studies on locust rectum: III Stimulation of electrogenic chloride transport by hemolymph. Can J Zool **58**:1933–1939

Spring JH, Hanrahan J, Phillips JE (1978) Hormonal control of chloride transport across locust rectum. Can J Zool **56**:1879–1882

Spring KR, Kimura G (1978) Chloride reabsorption by renal proximal tubules of *Necturus*. J Membr Biol **38**:233–254

Stewart WW (1978) Functional connections between cells as revealed by dye-coupling with a highly fluorescent naphtholamide tracer. Cell **14**:741–759

Strausfeld NJ, Miller TA (eds) (1980) Neuroanatomical techniques: Insect nervous systems. Springer, New York

Stobbart RH, Shaw J (1974) Salt and water balance: Excretion. In: Rockstein M (ed) The physiology of insecta, 2nd ed, Vol 5, Academic Press, New York, pp 361–446

Sutcliffe DW (1962) The composition of hemolymph in aquatic insects. J Exp Biol **39**:325–343

Swaroop A (1973) Micromethod for the determination of urinary inorganic sulfate. Clin Chim Acta **46**:333–336

Thomas RC (1978) Ion-sensitive intracellular microelectrodes: How to make and use them. Academic Press, New York

Treherne JE, Buchan PB, Bennett RR (1975) Sodium activity of insect blood: Physiological significance and relevance to the design of physiological saline. J Exp Biol **62**:721–732

Turin L, Warner A (1977) Carbon dioxide reversibly abolishes ionic communication between cells of early amphibian embryo. Nature **270**:56–57

Uhlich E, Baldmus CA, Ullrich KJ (1968) Behaviour of CO_2-pressure and bicarbonate in the countercurrent system of renal medulla. Pflugers Arch **303**:31–48

Ullrich KJ, Frömter E, Baumann K (1969) Micropuncture and microanalysis in kidney physiology. In: Passow H, Stämpfli R (eds) Laboratory techniques in membrane biophysics. Springer, New York, pp 106–129

Ussing HH (1948) The active ion transport through the isolated frog skin in the light of tracer studies. Acta Physiol Scand **17**:1–37

Ussing HH (1949) The distinction by means of tracers between active transport and diffusion. Acta Physiol Scand **19**:43–56

Ussing HH, Zerahn K (1951) Active transport of sodium as the source of electric current in the short-circuited isolated frog skin. Acta Physiol Scand **23**:110–127

Vurek GG, Warnock DG, Corsey R (1975) Measurement of picomole amounts of carbon dioxide by calorimetry. Anal Chem **47**:765–767

Walker JL (1971) Ionic specific liquid ion exchanger microelectrodes. Anal Chem **43**:89A–93A

Wall BJ, Oschman JL (1970) Water and solute uptake by rectal pads of *Periplaneta americana*. Am J Physiol **218**:1208–1215

Wall BJ, Oschman JL (1975) Structure and function of the rectum in insect excretion. Fortschr Zool **23**:192–222

Walser M (1970) Role of edge damage in sodium permeability of toad bladder and a means of avoiding it. Am J Physiol **219**:252–255

Wang CH, Willis DL, Loveland WD (1975) Radiotracer methodology in the biological, environmental, and physical sciences. Prentice-Hall, Englewood Cliffs, NJ

Watlington CO, Smith TC, Huf EG (1970) Direct electrical currents in metabolizing epithelial membranes. Exp Physiol Biochem **3**:49–159

Weidler DJ, Sieck GC (1977) A study of ion binding in the hemolymph of *Periplaneta americana.* Comp Biochem Physiol **56A**:11–14

White JF (1977) Activity of chloride in absorptive cells of *Amphiuma* small intestine. Am J Physiol **232**:E553–E559

Williams D (1976) Ion transport and short-circuit current in the rectum of the desert locust *Schistocerca gregaria.* Master's thesis, University of British Columbia, Vancouver, Canada

Williams D, Phillips J, Prince W, Meredith J (1978) The source of short-circuit current across locust rectum. J Exp Biol **77**:107–122

Wills NK, Lewis SA, Eaton DC (1979) Active and passive properties of rabbit descending colon: A microelectrode and nystatin study. J Membr Biol **45**: 81–108

Windhager EE (1968) Micropuncture techniques and nephron function. Butterworths, London

Wood JL (1972) Some aspects of active potassium transport by the midgut of the silkworm, *Antheraea pernyi.* Doctoral dissertation, University of Cambridge, Cambridge, England

Wood JL, Harvey WR (1975) Active transport of potassium by the *Cecropia* midgut; Tracer kinetic theory and transport pool size. J Exp Biol **63**:301–311

Wood JL, Harvey WR (1976) Active transport of calcium across the isolated midgut of *Hyalophora cecropia.* J Exp Biol **65**:347–360

Wood JL, Moreton RB (1978) Refinements in the short-circuit technique, and its application to active potassium transport across the *Cecropia* midgut. J Exp Biol **77**:123–140

Wright P (1967) A simple device for electronic measurement of a short circuit current. J Physiol (Lond) **178**:1P–2P

Wyatt GR (1961) The biochemistry of insect haemolymph. Ann Rev Entomol **6**:75–102

Zerahn K (1980) Competition between potassium and rubidium ions for penetration of the midgut of *Hyalophora cecropia* larvae. J Exp Biol **86**:341–344

Zeuthen T (1980) How to make and use double-barreled ion-selective microelectrodes. In: Bronner F, Kleinzeller A (eds) Current topics in membrane and transport, Vol 13. Academic Press, New York, pp 31–47

Chapter 4

The Study of Atmospheric Water Absorption

J. Machin

I. Introduction

1. Techniques

Atmospheric absorption is one of the most unusual and intriguing examples of water transport in animals, involving mechanisms that move water against spectacular thermodynamic gradients (Machin 1979b). Physiological study of the phenomenon has required the development of a wholly different experimental technology from that normally found in the transport physiologist's laboratory. The experimenter must explore water movements in both gas and liquid phases, find ways of measuring biological fluids of exceedingly high solute concentration and correspondingly reduced solvent activities, and frequently overcome the difficulties of working with very small animals.

The understanding of how living organisms transport materials against their thermodynamic gradients depends on the identification, measurement, and correlation of the relevant driving forces with the observed fluxes. In the case of water, driving forces are concentration gradients or chemical potentials. The common practice of expressing water vapor concentration as vapor pressure, relative humidity, or activity will be used here. The activity of water (a_W) is simply relative humidity expressed in ratio form. It is also sometimes necessary to make comparisons between gas and liquid phases, where solute concentrations are usually expressed as osmotic pressures. Formulae for converting osmotic pressures to solvent activity based on the equivalence of depression of freezing point and vapor pressure lowering are given below. The alternative practice of expressing liquid phase water concentration as water potential has not to the present been used in vapor uptake studies with animals and will not therefore be adopted in this chapter.

In this century impressive advances have been made in our understanding of the biological transport of liquid water and of dissolved substances. This progress is due partly to the fact that whole organisms or their functional parts can easily be maintained alive for long periods in conditions of stability in aqueous solutions. Generations of physiologists have found that practically any preparation can be maintained in the laboratory without elaborate control equipment. All that is required to maintain sufficiently constant water and solute concentrations and stable temperatures are a few tens of milliliters of aerated Ringer's solution. In studies of water vapor transport it is equally desirable to place the experimental animal in stable conditions where its water vapor concentration is known with some certainty. However, the low water-holding capacity of air, its low specific heat, and its tendency to form stable inhomogeneties make the maintenance of known water vapor concentrations difficult without some ingenuity, usually involving elaborate controlling equipment. It is no accident that the majority of studies of evaporative water loss in the biological literature employ dry air (Loveridge 1980). Water removal is the easiest form of regulation. Unfortunately, this simple expedient cannot be used in the study of water vapor absorption; for the greatest accuracy, experiments must be performed in the upper range of subsaturated water vapor concentrations.

As in the case of other technique-intensive areas of science, technological innovation has been directly responsible for scientific advancement. The most obvious of these have been the development of humidity-measuring instrumentation, continuously recording electronic balances, and accurate temperature-regulating devices. Among the least obvious, but nevertheless important where the constant humidity apparatus is temperature regulated by immersion in water baths, are electrical heating tape and effective tube couplings, both readily available from several manufacturers. These devices are essential in conveying humidified air from one water bath to another without risk of condensation and in making watertight connections between plastic chambers and soft metal tubing such as copper (Machin 1976). Despite the increased complexity of apparatus required to effectively study water transport in the gas phase, there are a few methodological advantages over conventional liquid phase investigations.

Gravimetric techniques can be used to measure water fluxes rather than the less convenient or accurate volumetric methods. The production or absorption of latent heat as water is gained or lost from an animal also opens the way for sensitive, rapidly responding methods of temperature measurement to be used as indicators of water flux. Most animals absorb water vapor by means of fluid compartments of exceedingly high osmotic pressure (Machin et al. 1982); these provide a much greater range of water concentrations than is found in nonvapor absorbers. Useful information can be conveniently obtained about the distribution of these concentration differences with less sensitive melting point techniques than are usually necessary.

2. Animals

Vapor absorption from the atmosphere is "a mechanism by which specially adapted terrestrial species, without access to water sources of high activity, are able to expend energy to escape from the conventional equilibrium dictated by the difference between environmental and body fluid water activities . . . ." (Machin et al. 1982). Species exhibiting this remarkable ability are to be found among insects and acarines inhabiting xeric environments or depending on intermittent food supply, such as blood feeders (Edney 1977). Most species must be in a state of water deficit before they begin to absorb vapor. Workers joining the field may be encouraged to find that more and more species capable of absorbing water vapor are being discovered. For the names of known vapor absorbers, including some little-known examples, consult Knülle and Spadafora (1970), Rudolph and Knülle (1982), Rudolph (1982a), and Wharton and Arlian (1972). It is also becoming clear that several, if not many, quite distinct mechanisms are involved, each representing different technical and physiological challenges.

Knowledge of current scientific ideas relating to vapor absorption and of their progression from information about the whole animal to component parts of the absorption mechanism may greatly aid future investigations. Previous investigations have progressed from the recognition that vapor uptake takes place in a particular species, to the identification of the site where it takes place. Initial rates of water uptake are net values and had to be resolved into separate gains and losses. This separation enables the particular characteristics of the uptake mechanism itself to be defined. The study of uptake kinetics gave valuable information about the nature of the driving forces and their maintenance when the pump is operating at different rates. The threshold of absorption provided a measure of the water concentration within a special fluid compartment or sink maintained by the absorbing animal. Identification of the compartment, within the animal or on its surface, by its markedly lowered water activities led to a more precise identification of the structures involved. In several instances it was possible to isolate important elements of the vapor-absorbing mechanisms and study their properties. These studies provided insights into the nature of activity lowering and the means by which water is conducted from one element of the mechanism to another.

II. Whole Animal Exchange Kinetics

1. Principles of Experimental Chamber Design

A. Temperature Control

Unlike water loss studies, investigations of water vapor uptake must be performed in high, stable, accurately known humidities or water activities. It cannot be stated too strongly that the stability of water concentrations in air depends directly on the precision of temperature control. Estimates of the errors in humidity control due to temperature fluctuations should be made when assessing the performance and working range of new apparatus. Detailed calculations are most conveniently made using an empirical equation relating temperature and saturation vapor pressure in List (1958) or the *Handbook of Physics and Chemistry* (The Chemical Rubber Company, Cleveland, Ohio). The best least squares fit over a narrow temperature range can be obtained by an exponential equation of the form

$$\text{Saturation vapor pressure} = a \cdot e^{b \cdot \text{temperature}} \tag{4.1}$$

or over wider temperature ranges with more complex equations (see Richards 1971). My own calculations indicate that variation in activity is almost directly proportional to temperature fluctuation, of the order of 6% per degree, with very little dependence on the nominal activity or temperature. For example, if the temperature of the chamber fluctuates ±1° about 25°C, the nominal saturated vapor pressure of 23.76 mm Hg may be as high as 25.21 or as low as 22.38 mm Hg. Under these conditions, air with a water activity of 0.75, for example, may actually be as low as

$$\frac{23.76 \times 0.75}{25.21} = 0.71\ a_w \text{ at } 26°\text{C}$$

and

$$\frac{23.76 \times 0.75}{22.38} = 0.80\ a_w \text{ at } 24°\text{C}$$

It is desirable in some instances to study vapor absorption at the highest possible activities. Temperature–vapor pressure relationships shed light on two related problems. Condensation on the chamber walls is highly undesirable because uncontrolled losses of water from the vapor phase, or contributions to it by evaporation of previously condensed droplets, seriously destabilize humidity within the experimental chamber. Condensed droplets on the experimental animal also introduce the possibility of nonvapor uptake by drinking or by osmosis through the body wall. Activities for incipient condensation at different levels of temperature control are given in Table 4.1; they are independent of the nominal temperature value. Passive osmotic flow into the animal begins when ambient activity exceeds that of the blood at activities of 0.987 to 0.995 for terrestrial arthropods having hemolymph concentrations varying from 0.3 to 0.7 osmol kg^{-1}. These figures indicate that more precise temperature control than ±0.1°C is unnecessary because the highest usable humidity is limited beyond this point by the activity of the blood, not the precision of the temperature regulation. Fluctuations in ambient activity of ±0.6% at this level of temperature regulation are acceptable.

In my experience direct temperature control to ±0.1°C of air surrounding any apparatus, while permitting access for the experimenter, is impractical if not impossible. By contrast, the desired accuracy of temperature control is within

Table 4.1. Activities for Incipient Condensation for Different Levels of Temperature Control

Temperature fluctuation (±°C)	a_w
0.1	0.994
0.2	0.988
0.4	0.977
0.6	0.964
0.8	0.952
1.0	0.940

the capability of the best modern water- or refrigerator-cooled circulating water baths. There is some loss in the precision of control when, as usually is the case, the apparatus has to be immersed in a second water bath indirectly controlled by the circulator. However, accuracy of control can be regained, at least partially, by housing the entire apparatus in an insulated or temperature-controlled room or by increasing the thermal inertia of the apparatus by incorporating blocks of copper or aluminum into its structure. Where increased accessibility can be exchanged for some loss in the accuracy of temperature control, the apparatus can be regulated by insulating circulating water jackets rather than by full immersion. The same level of temperature control, but with greater expense, could be achieved by constructing experimental chambers from large directly heated metal blocks insulated on the outside. This method is not yet in widespread use in vapor uptake studies.

B. Closed Chambers

It is difficult to provide any theoretical guidelines on the design of self-contained or closed humidity chambers because of complexities of convective circulation brought about by local temperature and humidity differences. The important guiding principles relating to closed-chamber design are that evaporative cooling and slowness of diffusion generally restrict the mass of humidified air close to the source of water. Equilibration in air spaces more than a few millimeters from the source takes many days if not weeks to be established. Self-contained chambers consist of a humidity-controlling solution at the bottom, a perforated floor above the solution, and a lid with access points for experimental animals and instruments.

Humidity-controlling solutions used to employ strong acids or alkalis (Solomon 1951), however, saturated salt solutions are highly preferable since they are less corrosive, are temperature insensitive, and tend to compensate automatically for water gained or lost. Winston and Bates (1960) provide the most complete information about a large number of salts as well as a fund of practical information about the technique. Since saturated salts at equilibrium give known stable activities, independent humidity measurement may not be required. To be sure the air in the chamber has a reasonable chance of being in equilibrium with the controlling solution, keep its surface area large and diffusion distances at a few millimeters. Petri dishes fitted with perforated floors therefore make excellent humidity chambers. At least 1 day must be allowed for equilibration before introducing the animal. This should be done through a small port, not by removing the lid. Larger, taller chambers can be used only when there is forced air circulating with a fan and when the prevailing humidity is checked by an independent measurement.

Any form of humidity-measuring probe may be used in still air. There are several electronic humidity probes of various sizes that could be used as long as they are small in relation to the experimental chamber. The best of these seem to be a new generation of semiconductor humidity sensors that are available with and without precalibration from several manufacturers. Alternatively, the humidity in enclosed spaces may be determined by weighing small pieces of hygroscopic material such as gelatin (Ludwig and Anderson 1942). Volume changes in drops of electrolyte solution (Dessens 1946; Weatherley 1960) may also be used (discussed in more detail in Sect. IV.2.B). There is also no reason that evaporation from a small droplet of pure water may not also be used, as long as the source of water is small in relation to the volume of the chamber and has a sufficient area of saturated salt solution to absorb it.

Gravimetric or psychrometric determinations have both been successfully used in my laboratory. The problem of keeping the surface area of the drop constant is easily overcome by holding it in a small area of filter paper, hanging it in a wire loop, or enclosing it in a short length of glass capillary. Calibration of any of these devices is best performed in a series of small sealed chambers lined with filter paper wicks and containing a series of saturated salt solutions. Winston and Bates (1960) also suggest the use of individual crystals of a series of salts for humidity measurement in very small chambers. This method has been exploited by Noble-Nesbitt (1975) in vapor uptake studies with *Thermobia*. Those salts whose solution vapor pressures are below the prevailing humidity will deliquesce while the remainder remain dry. Alternatively, when the crystals are moistened only the first group remains moist when the others become dry again.

Since humid air over a saturated salt solution collects at the bottom of a column of still air, humidity chambers under some conditions may be used with the top open. This permits access of external weighing devices or other instrumentation (Fig. 4.1). The opening must be restricted as much as possible to reduce the disturbing effects of external air currents. The introduction of filter paper wicks around the chamber walls deepens the zone of humid air above the solution and reduces the discrepancy between the observed humidity and that expected from the saturated salt employed. Presumably, wicks work by establishing convective mixing of air within the chambers and are therefore recommended whenever humidifying solutions are employed.

Since a vertical humidity gradient must exist in this system, even with wicks,

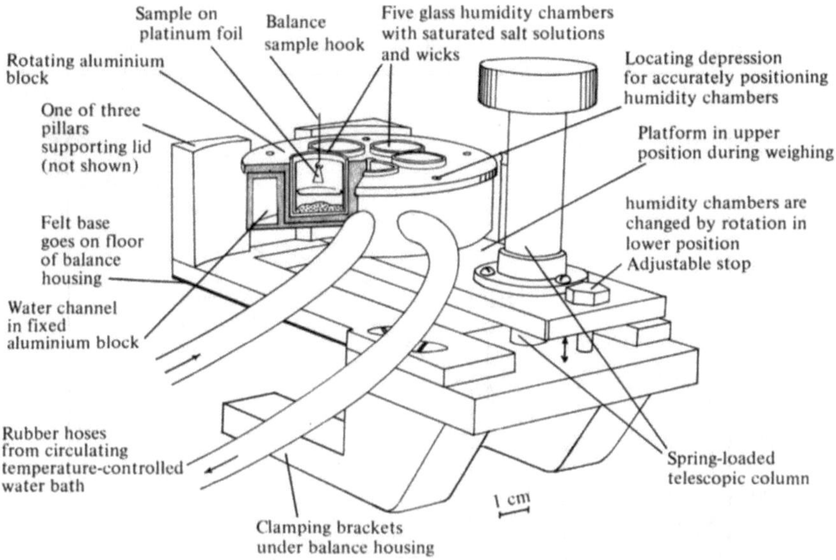

Fig. 4.1. Humidity-regulating insert used in conjunction with a Mettler ME22 electronic microbalance. One of five different humidities can be positioned around the sample hook. Humidity chambers contain saturated salt solutions and wicks. The chamber in use remains open to the balance mechanism by a small opening around the hook; those not in use are completely sealed by a lid (not shown). The temperature-controlling part of the apparatus has been cut away. [From Machin J (1979a) Compartmental osmotic pressures in the rectal complex of *Tenebrio* larvae: Evidence for a single tubular pumping site. J Exp Biol **82**:123–137]

it is important to determine the humidity exactly at the level occupied by the object under study. This can be done by measuring rates of weight change in hanging water drops (Machin 1979a). The expected rate of evaporation, assuming no gradients in the humidity chambers, is described by the heavy solid line in Fig. 4.2 drawn from the evaporation rate in ambient air at the graph origin. It can be seen that the discrepancies between observed rates of evaporation and the expected are considerably reduced in the presence of the wicks. The horizontal displacement between pairs of arrows indicates the difference between the humidity at the surface of the solution and that at the level of the drops.

C. Flow-Through Systems

Experiments involving continuous weight recordings or manipulation of the experimental animal frequently require larger experimental chambers than is feasible with closed systems. In this case the problem of humidity control is efficiently solved by slowly circulating air through the experimental chamber.

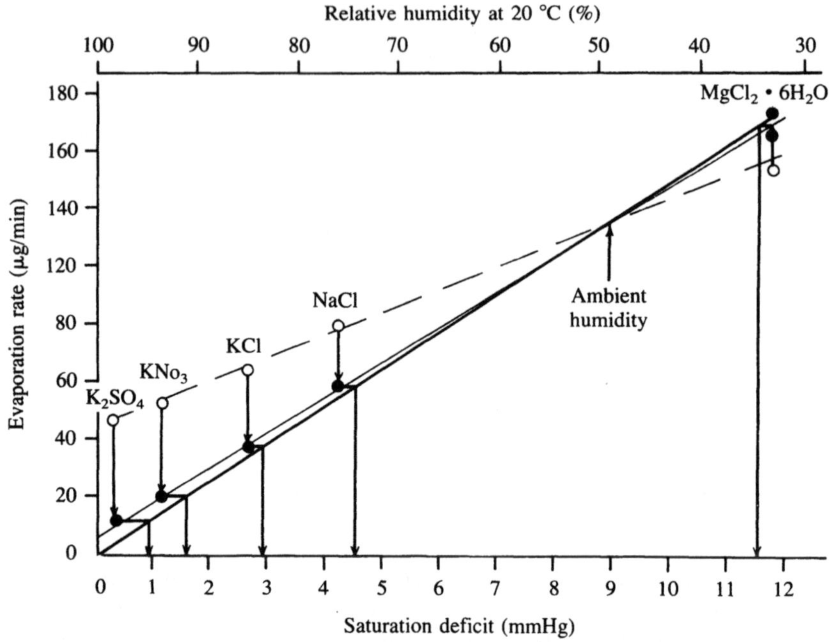

Fig. 4.2. Calibration graph of microbalance humidity chambers based on evaporation characteristics of a water drop without wicks (open circles, broken line) and with wicks (filled circles). The two continuous lines indicate the discrepancy with wicks between observed (fine line) and ideal evaporation rates (heavy line drawn between points at saturated and ambient humidities). The necessary corrections by interpolation to obtain the true humidity in each chamber are indicated by downward-pointing arrows. The saturated salts used in the chambers are also shown. [From Machin J (1979a) Compartmental osmotic pressures in the rectal complex of *Tenebrio* larvae: Evidence for a single tubular pumping site. J Exp Biol **82**:123–137]

The choice of flow rate is a compromise between the speed of humidity change and mechanical disturbance of air flow on the weighing. The technique has the added advantage of widening the range of usable control and measuring techniques. Various flow-through systems, including one I have developed, are described in more detail elsewhere (Loveridge 1980; Machin 1976). Some special features of this system are worth mentioning here. One requirement was the need to change the humidity rapidly from one value to another. It was found that enclosing the weighing mechanism in an airtight chamber connected to the weighing chamber underneath by a long, narrow "hang down" tube efficiently isolated it from the flow-through sections of the apparatus (Fig. 4.3a). Thus the total volume of the chamber, approximately 2 liters, was reduced to about 50 ml exchangeable air volume, greatly reducing the time required for humidity change.

It is desirable, for ease of construction, strength, lightness, and visibility, to construct experimental chambers of transparent plastic. All plastics are measurably hygroscopic and slowly exchange water with their surroundings. Although the rates of exchange by the chamber walls are not fast enough to measurably affect humidity control in a flowing system, hygroscopic absorption would seriously interfere with the actual weighing. To avoid this the weighing system, including animal cages, must be of metal. It is extremely important when the apparatus is immersed in water baths for temperature regulation that access ports and tube coupling be completely waterproof. The slightest leak, even though there may be no liquid entry, saturates the air stream. Access ports can be made to be readily removable by using screw threads, and easily sealed with silicone-greased neoprene O-rings. Beveled lips on removable ports or chambers force the O-ring into its seating to affect the seal (Fig. 4.3a). For connections between tubing and chambers, highly effective proprietary vacuum-tight tube couplings are ideal.

For a regulated supply of humidified air to the chambers, saturated salt solutions are unworkable because the precipitation of salt on the aerators interferes with the air flow. Bubble equilibrators require too long a water column for efficient temperature regulation. There are two effective methods: The first involves mixing regulated saturated and dried air streams in different proportions. Flow regulation is best performed upstream, before humidification, with paired float-type tube flow meters equipped with needle valves. The proportional flow through each tube is varied, keeping the total flow constant.

In the second method humidified air is brought to a lower well-regulated temperature (±0.01°C) by being passed through a heat-exchanger apparatus composed of copper slabs bolted together with 3 mm air spaces between. Details of the construction of the heat exchanger, which is immersed in its own temperature-regulated water bath, are given in Fig. 4.3b and c. Water condensed onto the copper is conveyed downward along shallow grooves to a reservoir, where it is intermittently removed. The temperature of the copper slabs determines the dew point of the emergent air, which is first equilibrated to final experimental temperature by being passed through a 3 m copper tube. Water vapor is prevented from condensing by copper tubes between water baths being heated with heating tape a few degrees above experimental temperature. Humidities may be rapidly changed from one value to another by using two heat exchangers in separate water baths. Air streams from either device are switched with a three-way manually or solenoid-operated valve.

Air flow through the apparatus is chosen so that uptake of water vapor cannot significantly affect the prevailing ambient humidity. With small insects of low permeability, flows of about 100 ml min^{-1} can be used. In flow-through systems, chamber humidity may be determined by any form of humidity sensor mounted in the emergent air stream. For routine work I use a mirror-type electronic dew point hygrometer, whose sensitivity is adequate over the entire humidity range but is particularly sensitive at low dew points. Several types of

(a)

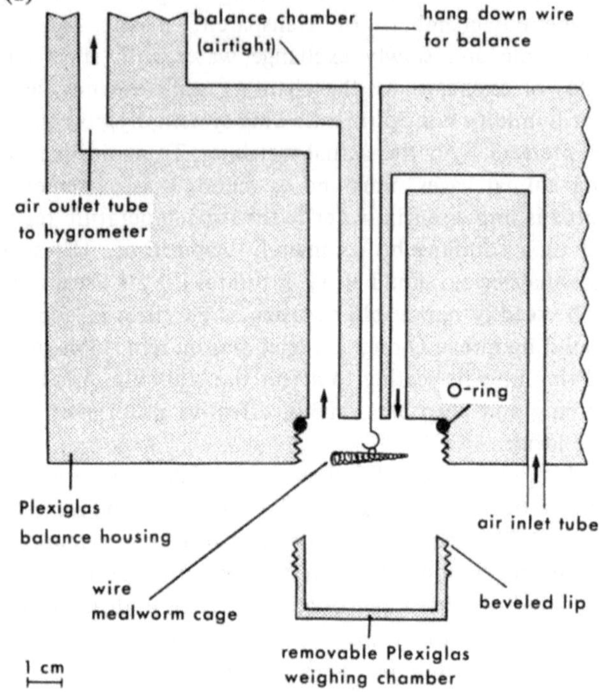

(c)

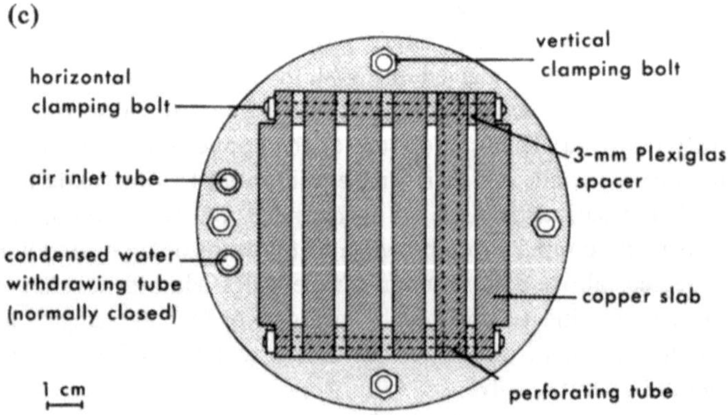

Fig. 4.3. Details of construction in flow-through apparatus used in the study of water vapor absorption. **(a)** Section of weighing chamber showing lay out of "hang down" tube and air inlet and outlet tubes. The outlet tube has a larger cross section to minimize pressure increase in the chamber. Arrangement of weighing chamber threading and O-ring, which provides an efficient seal without compromising ease of chamber removal, are also shown. **(b), (c)** Partial vertical

(b)

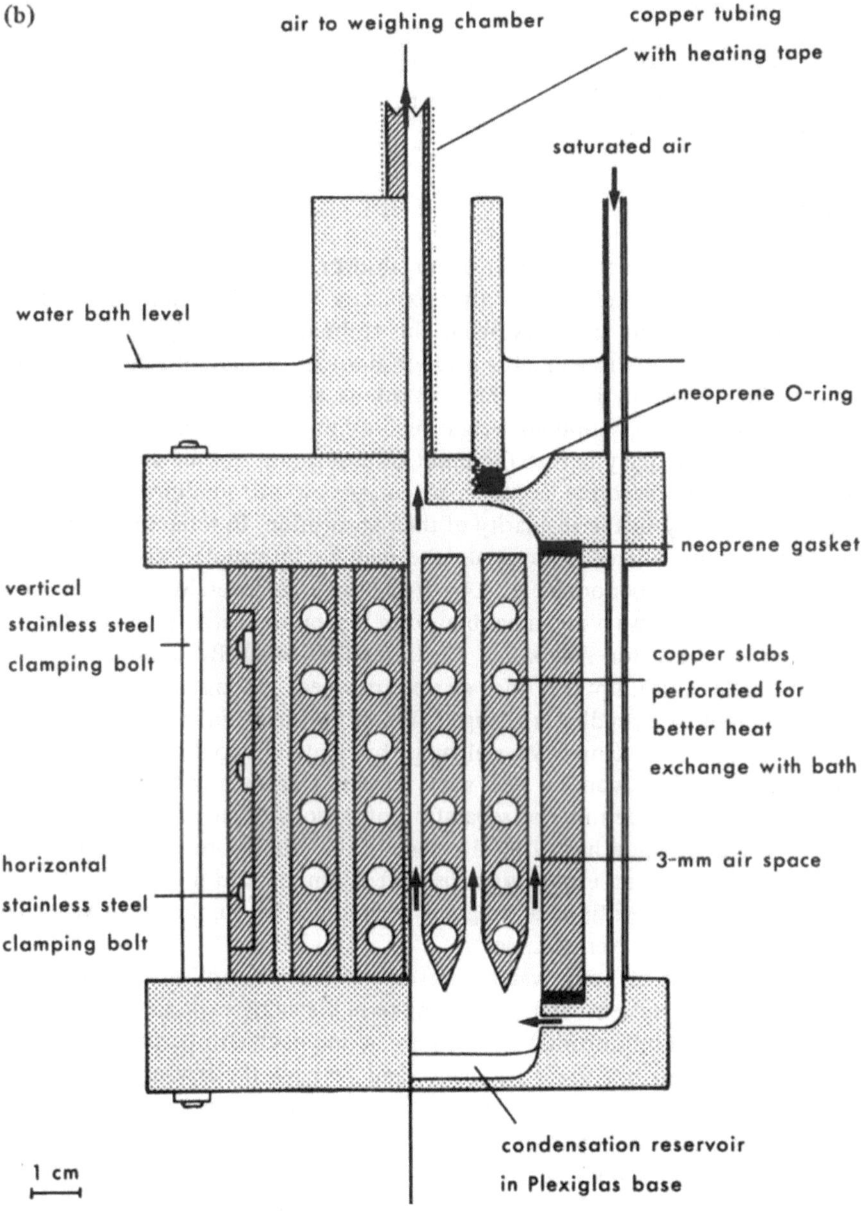

section and cross section of heat-exchanger apparatus used to provide air of regulated dew point. Construction materials are coded by shading. [Redrawn from Machin J (1976) Passive exchanges during water vapor absorption in meal-worms (*Tenebrio molitor*): A new approach to studying the phenomenon. J Exp Biol **65**:603–615]

semiconductor humidity sensors are now available and these would be equally suitable for general humidity measurement. For very high humidity studies a Wescor water potential probe (see Sect. IV.2.B) is the best instrument because the psychrometric method on which it is based provides an accurate reference to saturation point.

2. Balance Techniques

Much of the early work on water vapor absorption consisted of weighing the experimental animal before and after exposure to high humidity for periods varying between hours and days. While this technique is usually adequate to reveal whether or not the animal is a vapor absorber, it may not be sufficiently accurate to make physiological interpretations about the mechanisms. More recent studies using continuous weight recordings indicate that in most species there is a variable delay before the onset of absorption, presumably as the animal explores and becomes accustomed to its new surroundings. Species also differ considerably in the regularity of their absorption. In some species the pattern is highly irregular and vapor uptake may be interrupted at any time for short or prolonged periods (Fig. 4.4). It seems to be generally true that vapor uptake is in some way incompatible with locomotor activity (Noble-Nesbitt 1969), as the trace for *Thermobia* illustrates (Fig. 4.4). By contrast, in tenebrionid beetle larvae, uptake, once started, proceeds without interruption even when the ambient humidity is changed (Fig. 4.5). Clearly, the technique of calculating uptake rates from two weighings by assuming continuous absorption may be invalid when working with some species. Continuous weight recordings on the other hand, permit the separation of periods of sustained uptake from those where absorption has stopped or where the rate has become distorted by unrelated, intermittent events such as defecation or salivation. It is strongly recommended that continuous weighing techniques be employed in vapor absorption studies whenever possible.

The restoration of body water reserves by vapor absorption can, at least in the larger species, take several days. Long-term weighing experiments can yield a great deal of useful information but may be impossible to perform with some species. Mealworms, on the other hand, and perhaps the larvae of other tenebrionid beetles, are curiously suited to experiments lasting as long as several weeks. After an initial period of exploration, the larvae, perhaps because of their natural burrowing habitat, begin absorbing and lie completely quiescent in a closely fitting wire cage. Mealworms readily survive starvation. Lack of food in the gut drastically reduces the amount of defecation and excretion. After a few days without food, mealworms change their respiratory quotient from 1.15 to mean values of 0.73 (Johansson 1920), presumably because the metabolism switched from carbohydrate to primarily fat oxidation. Since starving mealworms therefore eliminate CO_2 at only 73% of the rate they consume oxygen, and since a molecule of oxygen is only 73% as heavy as a molecule of CO_2, the

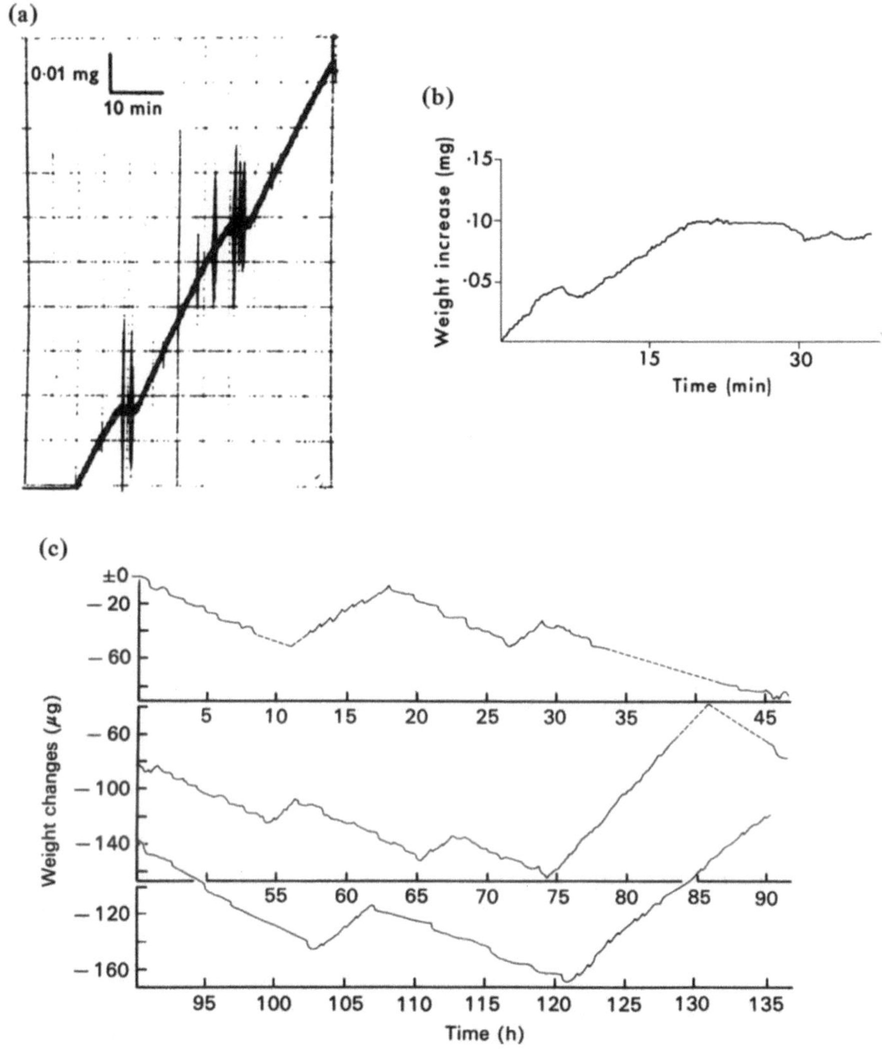

Fig. 4.4. Continuous weight traces during vapor absorption showing intermittent uptake of weight gain. **(a)** Vertical disturbances of the record obtained with *Thermobia domestica* indicate locomotory activity. [From Noble-Nesbitt J (1969) Water balance in the firebrat, *Thermobia domestica* (Packard). Exchanges of water with the atmosphere. J Exp Biol **50**:745–769] **(b)** *Arenivaga* sp. [From O'Donnell MJ (1980) Water vapor absorption by the desert burrowing cockroach *Arenivaga investigata* (Dictyoptera: Polyphagidae). Doctoral dissertation, University of Toronto] **(c)** *Amblyomma varigatum.* [From Rudolph DS, Knülle W (1978) Uptake of water vapor from the air: Process, site and mechanism in ticks. In: Bolis L, Schmidt-Nielsen K, Maddrell SHP (eds) Comparative physiology: Water, ions and fluid mechanics. Cambridge University Press, Cambridge, England]

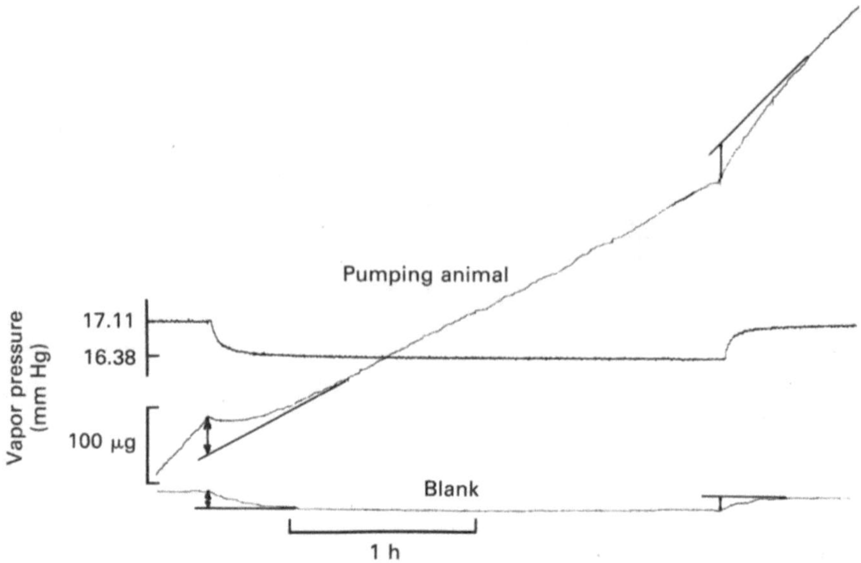

Fig. 4.5. Continuous weight traces during vapor absorption showing uninterrupted uptake pattern of *Tenebrio molitor* larvae. The effects of humidity change on uptake rate are also shown. Vertical arrows indicate non-steady-state weight gains and losses due to the humidity change. The blank trace indicates the extent to which these gains and losses are due to adsorption onto the empty weighing cage. [From Machin J (1976) Passive exchanges during water vapor absorption in mealworms (*Tenebrio molitor*): A new approach to studying the phenomenon. J Exp Biol **65**:603–615]

weights of metabolic gains and losses are balanced. Long-term weight losses, once corrections for occasional excretory elimination are made, must therefore provide accurate measurements of evaporation, without error from the consumption of food reserves in this animal. Several factors determine whether it is feasible or not to run long-term weighing experiments with a particular species. The most important consideration is one of metabolism and the magnitude of weight changes not due to water exchange.

Continuous weighing experiments can be designed to yield two basically different types of information. In constant humidities quasi-steady-state uptake rates can be measured. As long as the time taken to change humidity is short the initial physiological response to change will be under non-steady-state conditions. In mealworms (Machin 1976) there was a significant weight gain after the humidity was raised and a corresponding loss when it was lowered again (Fig. 4.5). There are two oppositely acting phenomena that introduce errors in such weight changes. Those due to density differences, buoyancy errors, are negligible. Surface adsorption, on the other hand, is significant in vapor absorption studies because the amount rapidly increases with humidity. The amount of adsorbed water can be generally predicted from blank experiments by constructing the weighing apparatus of nonporous metal and keeping it scrupulously clean. Cages should be cleaned after each experimental run in ultrasonicated detergent solution then rinsed with distilled water and then acetone, and then handled only with gloves. With this treatment adsorption errors for any given humidity change can be calculated using the knowledge that adsorption is inversely proportional to vapor pressure lowering. I have found that soiling of the metal weighing cages by the animal increased the adsorption error and its variability.

3. Tracer Methods

The measurement of tritiated water exchange is better suited to describing total body water kinetics than to identifying specific components of the exchange. Nevertheless, tracer techniques have been developed to study aspects of water balance in acarines (Wharton and Devine 1968), and in some cases attempts have been made to relate them to energy-dependent vapor uptake. Tracer methods have been used to study water exchange in individual house dust mites of the genus *Dermatophagoides* as small as 3.5 μg (Arlian 1975). Although they may be easier to perform with small animals than weighing experiments, they still depend on the gravimetric determination of total body water.

Tracer studies having relevance to vapor uptake involve measurement of fluxes in a series of ambient activities, preferably including 0 (dry air) and 1.0 (saturated air). Several measurements should be made in activities from which vapor uptake is possible, and others should be below the animal's environmental equilibrium, i.e., critical equilibrium activity (CEA). Influx measurements are performed by exposing groups of animals to tritiated water vapor in known activities. Individuals are then sacrificed at known time intervals and the body tritium is counted by liquid scintillation. For efflux, animals are tritium loaded for several days, then counted after timed exposure in tritium-free air. Since the sampling is intermittent, flux determinations suffer from the limitations of all such measurements in being unable to identify discontinuous exchange processes. In all animals studied the label is found only in the body water. Fluxes are submitted to conventional compartmental analysis using graphic techniques (Solomon 1960).

The principal difficulty of water tracer studies is that tritium influx reflects total tracer absorption. For reasons that are not understood, there is some difficulty in separating active and passive absorption because influx and efflux appear balanced. Furthermore, attempts to identify the active component to influx by CO_2 anesthesia have failed (Devine and Wharton 1973; Knülle and Devine 1972). An alternative approach has been possible with *Dermatophagoides farinae* (Arlian and Wharton 1974) because this species exhibits two compartment-exchange kinetics below CEA: one of small mass and short time constant, and the other larger with longer exchange times. Exchange above CEA is limited to the fast compartment. The slow compartment is assumed to represent passive exchange by the blood and the fast one the active vapor uptake mechanism. After slow exchanges above CEA are calculated by extrapolation, it is possible to obtain the vapor absorption component by subtracting the slow from the fast influxes. There is uncertainty as to whether this method is generally applicable to all vapor absorbers. Consult Arlian and Wharton (1974) for a fuller description of their arguments and calculations.

4. Thermocouples

The use of thermal changes to quantify either evaporation or condensation normally would not be considered as an alternative to gravimetric or tracer techniques. Thermal contact of thermocouple beads with the hard surfaces of arthropods is imperfect, as is the relationship between evaporation rate and cuticle temperature depression (Gilby 1980). However, thermocouple beads are small and they can be effectively used to pinpoint local sites of exchange by identifying where warming by condensation as opposed to evaporative cooling is taking place (O'Donnell 1977). The rapidity of response of thermocouples, perhaps in conjunction with direct observation, could be further exploited to distinguish between cyclical and continuous uptake processes.

5. Analysis of Kinetics

Even when the record of weight gain has been extracted from the "noise" of intermittent events, water exchanges between an animal and the environment must be separated into two principal components: general surface transpiration, and uptake by specialized condensing surfaces. The humidity at which uptake balances loss is the ecologically significant parameter, CEA. In some vapor absorbers, for example, *Tenebrio* larvae, measured weight gains directly reflect activities of the absorbing sink or pump because cuticular losses are insignificant (Fig. 4.6). In others, for example, *Arenivaga,* some *Amblyomma* species, as well as some of the smaller insects *Xenopsylla* and acarines *Acarus* with more permeable cuticles, surface losses must be added to measured weight changes to obtain the true uptake rate (Machin 1979b). Continuous weight records for *Amblyomma* (Fig. 4.4c) illustrate this point. It can be seen that the slopes of the periods of net loss are of similar magnitude to those of net gain.

In order to calculate the vapor uptake rate due to the pump, cuticular losses must be added to the net weight change of the animal. These losses can either be determined from continuous weight records when the pump is not working, or can be obtained in humidities too low for vapor uptake. In the case of *Tenebrio* (Fig. 4.6), information about water loss kinetics from the absorbing compartment could be obtained because the anus remained open in vapor-absorbing animals suddenly subjected to subthreshold humidities. The same kind of passive loss from a separate absorbing compartment was apparent in the tritium-exchange kinetics of *Dermatophagoides* (Arlian and Wharton 1974).

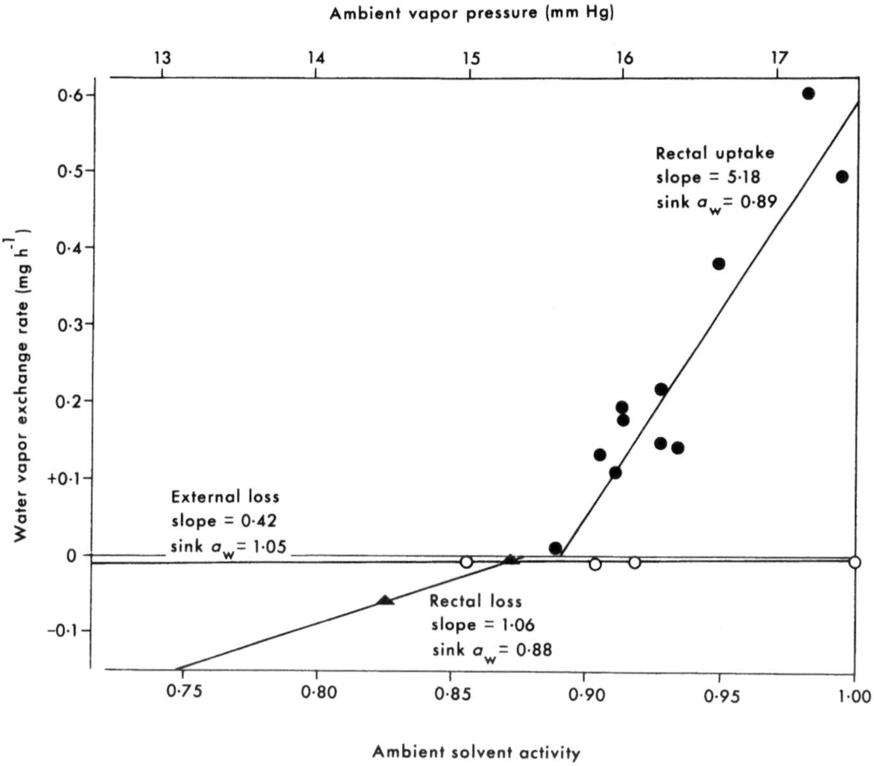

Fig. 4.6. Water-exchange compartments of *Tenebrio molitor* larvae obtained by gravimetric methods under quasi-steady-state conditions. Sink activities for each compartment were obtained by extrapolation.

Once appropriate corrections for surface losses have been made, uptake rates indicate the true performance of the mechanism. Several vapor-absorbing species exhibit tolerably linear uptake kinetics, in which the uptake rate increases proportionally with ambient activities or vapor pressures exceeding a threshold value. The simplest model explaining this is that water vapor is absorbed into a compartment of significantly lower activity, which is maintained at a constant level independent of the rate of uptake. By contrast, some species, especially the thysanurans *Thermobia* and *Ctenolepisma* (Beament et al. 1964; Edney 1971), exhibit markedly nonlinear uptake kinetics, suggesting that the absorbing compartment may become diluted at high uptake rates. In this case linear extrapolation is inappropriate and the threshold can only be determined directly by noting the humidity at which uptake ceases.

The analysis of non-steady-state weight changes has confirmed the existence of a sizable, passively exchanging fluid compartment associated with the vapor absorption mechanism in tenebrionid beetle larvae. The fact that compartments of significant size are not apparent in all vapor absorbers (e.g., *Thermobia*) suggests that non-steady-state experiments could be used to distinguish between uptake mechanisms in the future. The weight changes shown by mealworms following humidity change were first corrected for adsorption. The hypothetical compartment showing these weight changes appeared to have the normal properties of a solution in which total solvent volume varies inversely with the vapor pressure lowering (a measure of solute concentration). The total solvent volume (V_1) in either of two vapor pressures (vp_1 or vp_2) can be determined from the following equation (Machin 1976, 1978):

$$V_1 = \frac{\delta V \cdot vp_0 - vp_2}{vp_1 - vp_2} \tag{4.2}$$

where vp_0 is the saturation vapor pressure and δV is the volume change between the two humidities, determined by the weight change (solvent weight in milligrams can be directly converted to volume in microliters).

III. Site of Absorption

1. Occlusion Experiments

The discovery by Noble-Nesbitt (1970a) that absorption occurred in areas of the body surface and involved special structures was an important turning point in our understanding of vapor transport. The most widely used method of occluding different areas of the body in order to inhibit absorption has been the application of molten wax, which subsequently solidifes to form an obstruction to vapor uptake (Machin 1975; Noble-Nesbitt 1970a, 1970b; O'Donnell 1977; Rudolph, 1982b; Rudolph and Knülle 1974, 1978). Dunbar and Winston (1975) ligatured the abdomen to interfere with rectal uptake in *Tenebrio,* perhaps with more ambiguous results. Quite correctly, Okasha (1971) has criticized the wax application technique on the grounds that the wax might only interfere with vapor uptake indirectly, by heat injury or sensory interference. In my experience, materials such as beeswax and paraffin wax are brittle, tend to crack, and eventually separate from the cuticle. The wax technique can be improved by using a mixture of 10 g beeswax and 4.5 g colophony (Krogh and Weis-Fogh 1951), a preparation that has a lower melting point and is stickier, softer, and less prone to cracking.

2. Reversible Exposure

Workers have answered Okasha's criticism by the "reversible exposure" technique, independently developed by Rudolph and Knülle (1974) and Noble-Nesbitt (1975) and subsequently employed by O'Donnell (1977). The animal is mounted with the body between separate anterior and posterior chambers (Fig. 4.7). The ability of the animal to absorb water vapor when each end of the body, in turn, is exposed to humid air is detected by weight changes in the animal or by volume changes in salt solution droplets within either chamber.

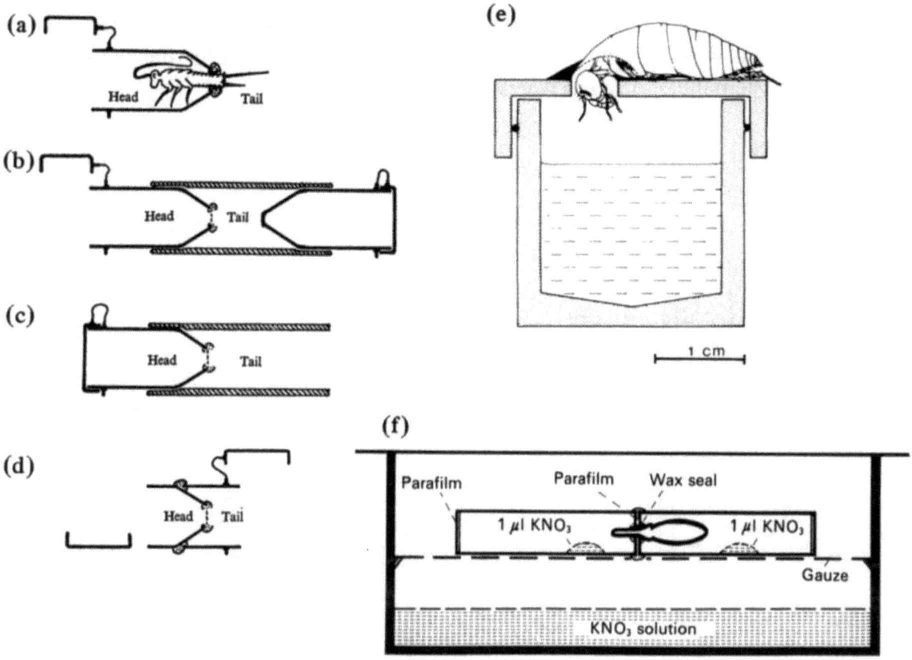

Fig. 4.7. Reversible exposure techniques employed by different workers. **(a)–(d)** Techniques used for *Thermobia domestica* in which the head or tail end of the animal can be selectively exposed to any given humidity. [From Noble-Nesbitt J (1975) Reversible arrest of uptake of water from subsaturated atmospheres by the firebrat, *Thermobia domestica* (Packard). J Exp Biol **62**:657–669] **(e)** Apparatus used for *Arenivaga* sp. consisting of an O-ring–sealed stainless steel chamber containing saturated salt solution. The lid and wax-sealed animal can be removed for weighing. [From O'Donnell MJ (1977) Site of water vapor absorption in the desert cockroach, *Arenivaga investigata.* Proc Natl Acad Sci USA **74**:1757–1760] **(f)** Apparatus used for *Amblyomma varigatum* in which two 1-μl drops of saturated KNO_3 are introduced through the Parafilm with a syringe and afterward sealed. The ability of the animal to absorb vapor at activities of 0.93 is recorded by measuring volume changes in either drop. [From Rudolph D, Knülle W (1978) Uptake of water vapor from the air: Process, site and mechanism in ticks. In: Bolis L, Schmidt-Nielsen K, Maddrell SHP (eds) Comparative physiology: Water, ions and fluid mechanics. Cambridge University Press, Cambridge, England]

3. Fecal Equilibration

In addition to unequivocally distinguishing between rectal and non-rectal uptake mechanisms, the fecal equilibration technique provides information on the mode of operation of the rectum in absorbing water vapor. It was originally developed by Ramsay (1964) to study the capacity of the rectal complex of *Tenebrio* to dehydrate the feces. With this ingenious approach freshly eliminated fecal pellets could be rapidly weighed and subsequently equilibrated to different known humidities (Fig. 4.8a and b). The humidity in the rectum could then be determined from the pellets' original weight by interpolation.

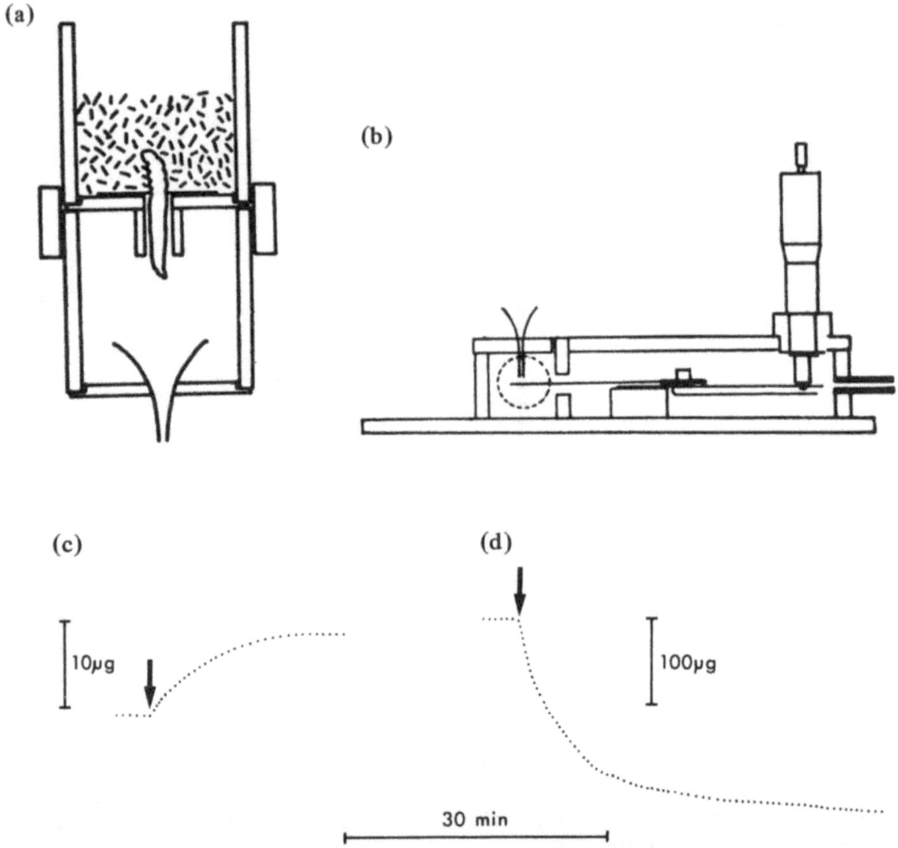

Fig. 4.8. Fecal equilibration techniques to determine whether the rectum is the site of vapor absorption. **(a), (b)** Feeding and weighing apparatus used for *Tenebrio molitor* larvae. Fecal pellets fall through the funnel and are stuck on the quartz fiber beneath for recording of their fresh weight and subsequent equilibration in known humidities. [From Ramsay JA (1964) The rectal complex of the mealworm *Tenebrio molitor* L. (Coleoptera, Tenebrionidae). Philos Trans R Soc Lond (Biol) **248**:279–314] **(c)** Weight trace of a non-vapor-absorbing mealworm in high humidity after it has produced a fecal pellet (arrow) retained by the weighing cage. **(d)** Weight trace of an *Arenivaga* sp. in high humidity after it has produced a fecal pellet (arrow) retained by the weighing cage. [Redrawn from O'Donnell MJ (1980) Water vapor absorption by the desert burrowing cockroach *Arenivaga investigata* (Dictyoptera: Polyphagidae). Doctoral dissertation, University of Toronto]

Essentially the same experiment can be performed more simply on a continuously recording balance by using recently fed experimental animals and covering the floor of the weighing cage with foil to prevent fecal pellets from being lost. Any change in water content of fecal pellets following elimination from the rectum can usually be identified in a continuous weight record. Fecal pellets in nonabsorbing mealworms were observed to gain weight in humidities exceeding the absorption threshold (Fig. 4.8c). This is the same result as Ramsay (1964) obtained, which demonstrated that the rectal contents, with the anus closed, became equilibrated to the low activities of the absorbing compartment within the animal. By contrast, fecal pellets produced by vapor-absorbing animals with the anus open made no further weight changes after elimination, giving a clear indication of unimpeded flow of water vapor into the rectum (Machin 1978). O'Donnell (1977), using the same technique, proved that the rectum could not be the site of uptake in *Arenivaga* because this animal's fecal pellets lost weight in humidities suitable for absorption (Fig. 4.8d).

IV. Physiology of Absorption Mechanisms

1. Morphometry

Morphometric analysis of vapor-absorbing structures can provide important insights into the mechanism in several different ways. The coefficient of diffusion of water vapor in air is a well-established quantity (Leighly 1937; Mason and Monchick 1965; Schwertz and Brow 1951) and can be used to test whether or not diffusion is rapid enough to account for the observed flux. This exercise is performed by incorporating the appropriate areas and diffusion distances, measured on the animal, into the calculation. Similar analyses can be extended to tissues within the body through which the condensed water has to pass. Such studies acquire added significance if morphometric data can be collected from a range of different-sized individuals or from several related species.

The dimensions of vapor-absorbing structures play a different role in uptake in *Arenivaga*. Low water activity in this animal depends not on high solute concentrations but on the maintenance of extremely high surface tensions of fluid between fine hydrophilic cuticular hairs. Since the diameter of these hairs bears a direct relation to the radius of curvature of the fluid surface, measurements of the hairs (O'Donnell 1981a, 1981b, 1981c, 1982) plays an important part in interpreting how they function. Scanning electron microscopy is an invaluable tool for making such measurements and elucidating the interrelationships of other surface structures that are an integral part of the uptake mechanism in *Arenivaga* and perhaps other species. Microstructures involved in interacting with water should always be dehydrated by critical point freeze drying in order to minimize distortion due to surface tension.

2. Properties of the Structures

Identification of the site of absorption permits detailed examination of the physiological properties of the structures involved. Kinetic studies with the intact animal point to the key involvement of an absorbing fluid of markedly lower activity than the rest of the body fluids. At the present state of our knowledge, it seems probable that in the majority of cases reduced activity is due to extremely high concentrations of various solutes. In some examples electrolytes have been shown to be the principle solutes, in others activity lowering may be brought about by soluble organic compounds. Recently, O'Donnell (1981c, 1982) showed that the desert cockroach *Arenivaga* stands apart from other known vapor absorbers in generating fluids of low activity without the aid of concentrated solutes. In addition to measuring osmotic pressures and solute concentrations directly, O'Donnell pointed out that there was a simple technique for unequivocally distinguishing between solute-coupled and solute-independent vapor-absorbing mechanisms based on the behavior of the absorbing fluid in high humidity. Once the flow of this fluid is interrupted or the fluid itself is removed from the animal, samples with high solute concentrations would continue to absorb water until equilibrated with the absorbing humidity, whereas fluids whose activity was lowered by some other means would dry out.

A. Analysis of Sections

One important consequence of low activity and high osmotic pressure of the fluid is that it and associated structures can be readily distinguished from hemolymph using the melting characteristics of frozen sections. It has been possible, using a compound microscope enclosed in the cold chamber of a freezing microtome fitted with a temperature-controlled stage, to photograph frozen sections mounted in kerosene in the process of melting at known temperatures (Machin 1979a). The distribution of osmotic pressures within the section, determined from the melting points, can then be obtained from the series of photographs. The resolution of such a technique is not very great when steps of 1°C (equivalent to about 0.5 osmol kg^{-1}) are used. This is acceptable, however, because of the very large range of osmotic pressures in the mealworm rectum.

It should be emphasized that the validity of the method depends on rapid initial freezing of the tissues under study. Isopentane cooled to the point of freezing (−150°C) by liquid nitrogen is recommended. This method is preferable to using liquid nitrogen directly because the gas layer formed around the object hinders the transfer of heat during freezing. Uniform melting points in large fluid compartments such as the hemolymph provide a useful criterion of rapid freezing. Gradients with higher melting points near the center of the section are observed when the whole mealworm is frozen. When only the posterior end of the animal containing the rectum is used, the melting point differences in the hemolymph disappear. Presumably, this is due to the smaller mass of the sample and perhaps because the hemolymph is in direct contact with the coolant.

In favorable circumstances, when sufficient sections from the same vapor-absorbing individual can be collected, it is possible to make a three-dimensional reconstruction of the osmotic pressures within the absorbing organ. It should be noted also that the greatest difficulty can be experienced in obtaining good, whole frozen sections of some parts of the rectum because of their fragility. Sections of the most concentrated areas, where the amount of freezable water is considerably reduced, readily disintegrated on handling, confirming the extremely dehydrated state.

Analysis of frozen sections has one key advantage over micropuncture techniques in that it preserves the precise interrelationship between neighboring fluid compartments. At a higher resolution, electron microprobe X-ray analysis of ultra-thin frozen sections has been used to study vapor absorption (Noble-Nesbitt 1978) and might possibly be exploited further.

B. Isolated Samples

The components of vapor-absorbing systems sometimes retain some of their functional properties when removed from the animal. In the isolated state these properties can be more extensively explored, frequently with more precision and less ambiguity of interpretation. In removing fluid samples from vapor-absorbing animals some workers have mistakenly interpreted the hygroscopic behavior of these fluids as proof of their involvement in vapor uptake. It must be realized that all solutions, no matter how dilute, gain or lose water passively depending on the prevailing ambient humidity. Definitive proof that the fluid serves to absorb water from subsaturated atmospheres in the intact animal must rest on a measurement of the sample's activity, or the demonstration of water absorption, at the time of production.

Insights into the functioning of fluid compartments in the water-transporting process may be further gained by studying the water content–concentration relations of both liquid and solid samples taken from the site of vapor uptake. In some instances the use of cryoscopy to measure the osmotic pressures of very concentrated solutions may yield anomalous results. Ramsay (1964) found that mealworm perirectal fluid, which contains large amounts of protein, displayed unequal freezing and thawing points. Anomalous freezing is well known to cold-water fish physiologists, who have been able to demonstrate that certain glycoproteins in the blood act as natural antifreezing agents (Hew et al. 1981). Freezing and thawing depends on the water molecules having sufficient freedom of movement to align themselves in crystal formation, as well as the sample's ability to conduct heat. Both these processes may have been impaired in the viscous proteinaceous perirectal fluid of mealworms.

Since perirectal fluid samples studied at environmental temperatures failed to show any anomalous behavior, vapor pressure methods for measuring the colligative properties of concentrated solutions are preferable. One workable technique is based on measuring volume changes of fluid droplets in a series of

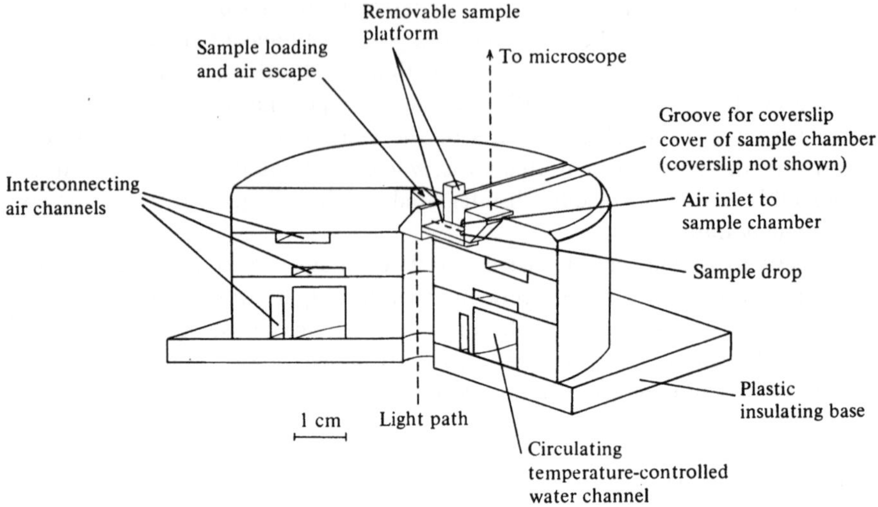

Fig. 4.9. Cutaway diagram of humidity chamber used to measure volume–concentration relationship of perirectal fluid. The apparatus is constructed of aluminum blocks bolted together with milled temperature-controlled water and air channels. The length and narrowness of the air channels (interconnection not shown) ensure that incoming air reaches the temperature of the blocks before passing over the sample drop. The prisms, which permit the sample to be viewed and illuminated from the side, together with the removable sample platform are drawn in full. The apparatus is mounted on a microscope stage with offset condenser and light source. [From Machin J (1979a) Compartmental osmotic pressures in the rectal complex of *Tenebrio* larvae: Evidence for a single tubular pumping site. J Exp Biol **82**:123–137]

standard humidities. The apparatus is a temperature-controlled humidity cell in which sample droplets (3–50 μl) can be viewed for the purposes of measurement with a compound microscope (Fig. 4.9). It consists essentially of an aluminum block through which water is circulated from an external water bath to regulate its temperature to 20.00° ± 0.01°C. Air streams of differing regulated vapor pressures are brought to this temperature before passing over the sample. Two 45°C prisms permit the sample to be illuminated and viewed from the side. The chamber holds up to five drops, including the KCl standard of known molarity of similar size and molar concentration to the unknowns. Chamber vapor pressures can be calculated from volume changes by using any electrolyte solution of known properties.

The concentrative properties of a large number of different salt solutions are available in the *Handbook of Physics and Chemistry* (The Chemical Rubber Company, Cleveland, Ohio). To obtain the vapor pressure calibration, first convert freezing point depression (δt) to vapor pressure lowering ($vp_0 - vp_1$ in millimeters mercury) using the following equation:

$$\mathrm{vp}_0 - \mathrm{vp}_1 = \frac{\delta t \times 18 \times \mathrm{vp}_0}{1.86 \times 1000} \tag{4.3}$$

then calculate by the least squares method the empirical relationships between the reciprocal of the molar concentration, M (liters of solution per gram mole) and $1/\mathrm{vp}_0 - \mathrm{vp}_1$.

For KCl, the following relationship fits a straight line very well ($r^2 = 1.00$):

$$\frac{\text{liters of solution}}{\text{g mol}} = \frac{-0.042 + 0.584}{\mathrm{vp}_0 - \mathrm{vp}_1} \tag{4.4}$$

For two different vapor pressures, vp_1 and vp_2,

$$\frac{\text{liters of solution}_2 \times \text{g mol}_1}{\text{g mol}_2 \times \text{liters of solution}_1} = \frac{\mathrm{vp}_0 - \mathrm{vp}_1}{\mathrm{vp}_0 - \mathrm{vp}_2} \tag{4.5}$$

When solution concentrations are changed only by adding or subtracting water,

$$\text{g mol}_1 = \text{g mol}_2 \tag{4.6}$$

and so cancel out.

Solving for $\mathrm{vp}_0 - \mathrm{vp}_1$ in Eq. (4.4) gives

$$\mathrm{vp}_0 - \mathrm{vp}_1 = (-0.042 + 0.584)M \tag{4.7}$$

Substituting in Eq. (4.5) yields

$$\frac{\text{liters of solution}_2}{\text{liters of solution}_1} = \frac{(-0.042 + 0.584)M}{\mathrm{vp}_0 - \mathrm{vp}_2} \tag{4.8}$$

where liters of solution$_2$/liters of solution$_1$ is the ratio of final to initial standard drop volume and M is the samples' initial molar concentration. Equation (4.8) can be solved to determine vp_2, the only unknown.

Both unknowns and calibration droplets are placed on a newly siliconed platform where they assume spherical shape. Their volumes are calculated from eyepiece micrometer measurements of height h and base diameter d using the following equation:

$$\text{Drop volume} = \pi h/6(h^2 + 3/2d^2) \tag{4.9}$$

In order to keep drop volumes and equilibration times (30-60 min for 10 nl drops) in unknowns and standards roughly similar, the initial molarity of the standard should approximately match that of the unknowns. Results obtained with this method using perirectal fluids of differing concentrations are illustrated in Figure 4.10. The use of the reciprocal of vapor pressure lowering on the x-axis linearizes the results. It can be seen that there is no trace of histeresis in the gains and losses of water by the drops.

Two further techniques available for studying the water content-concentration relationships of solid samples have been used for cuticle. The water content of most materials and solutes is strongly nonlinear, increasing rapidly at humidi-

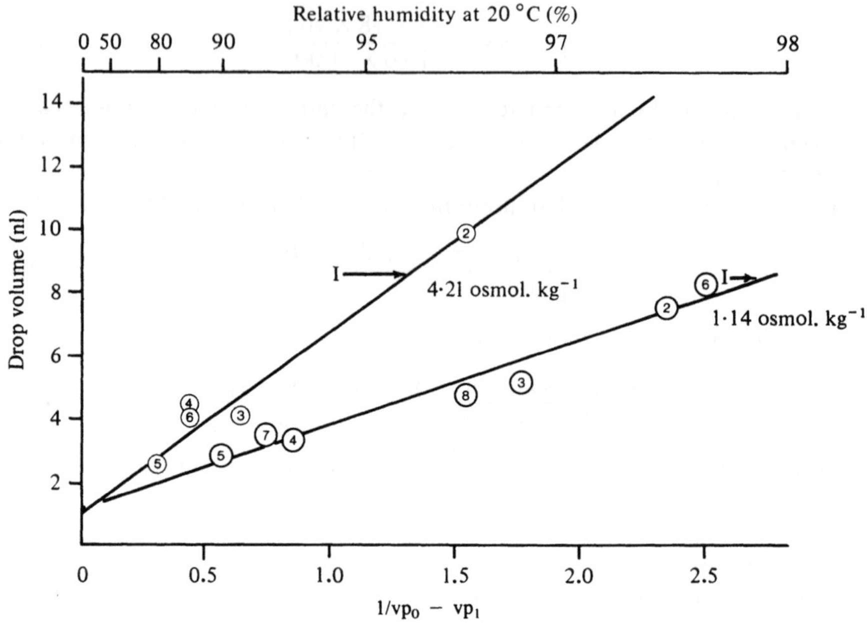

Fig. 4.10. Representative plots of drop volume versus the reciprocal of vapor pressure lowering of similar-sized samples of perirectal fluid from absorbing (fine open circles) and nonabsorbing (heavy open circles) mealworms. *I*, Initial volume of the drops and the osmotic pressure calculated by interpolation. The other points are equilibrium values measured in the order indicated by the numbering. [From Machin J (1979a) Compartmental osmotic pressures in the rectal complex of *Tenebrio* larvae: Evidence for a single tubular pumping site. J Exp Biol **82**:123–137]

ties close to saturation. For increased accuracy it is convenient to make measurements in very high humidities and extrapolate the results to lower values. The gravimetric method, suitable for sample weights of the order of 1 mm, utilizes a balance, preferably an electronic microbalance, fitted with some means of exposing the sample in a series of known ambient humidities (see Fig. 4.1).

The second method is suitable for smaller cuticle samples, down to a few micrograms, but is restricted to high ambient humidities (a_w = 0.920–0.996). The technique, devised by O'Donnell (1980, 1982) employs a water potential cell operated by a dew point microvoltmeter (Wescor, South Logan, Utah). The cell consists of a sealable air chamber, of the order to 40 μl in volume, containing a thermocouple. The instrument operates by an electronic time-sharing mechanism in which the thermocouple bead alternates between a cooling to dew point mode using the Peltier effect and a reading mode driving a meter to display the bead temperature (Campbell et al. 1973). A known amount of water is introduced into the chamber. At equilibrium it becomes distributed among the air, the thermocouple bead and chamber walls, and any liquid or solid samples present. The activity of all compartments is the same. Compared with adsorbed or condensed water the amount in the air is negligibly small, a maximum of 0.8 nl. Water content measurements are typically performed by placing a series of accurately known amounts of water into the chamber and recording the stable dew point with and without the sample present. The series without the sample, neglecting the vapor phase, yields the amount of adsorbed and condensed water at different activities (Fig. 4.11). Subtracting the amount of adsorbed and condensed water required to give the same activity in the presence of the sample gives the water content.

The precision of this method, when samples of a few micrograms are used, depends on the operator being able to introduce accurately known amounts of water varying between 5 and 200 μl without significant evaporation. The problem has been solved by using different areas of 0.2 μm pore size Millipore filter both to measure and deliver water into the chamber. Millipore filters, when dipped in water first, have a shining appearance but become dull when the superficial water is lost by evaporation. Using this end point to indicate precisely when superficial water is lost, water contents of known areas of filter are determined by using tritiated water. Known amounts of water are introduced into the

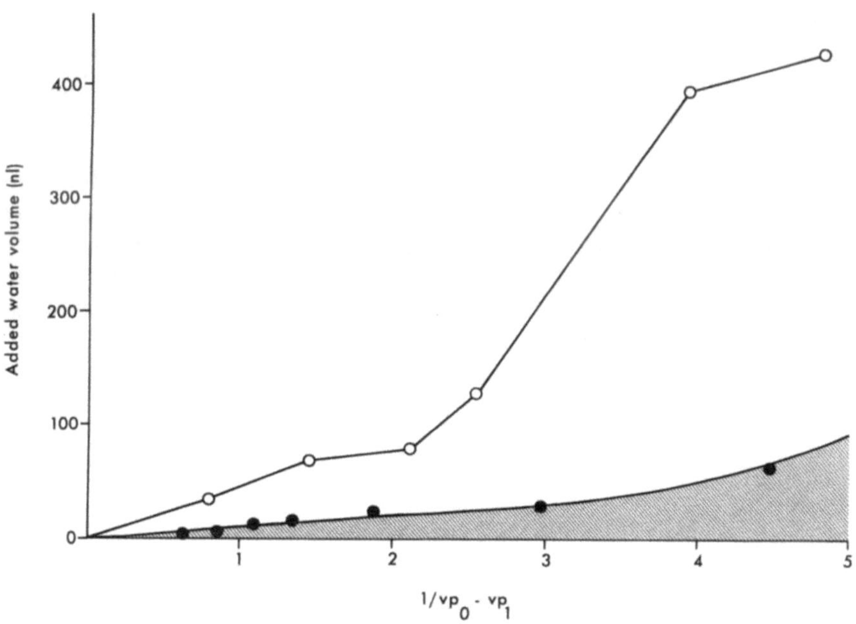

Fig. 4.11. Example of data obtained from the Wescor water potential cell at 22°C with 1.17 μg, 2.43 mm^2 *Arenivaga* cuticle (open circles) compared with an empty chamber (filled circles). The shaded area indicates the amount of water retained by condensation on the thermocouple bead or adsorbed on the chamber walls. Corresponding added water values of the blank must be subtracted to obtain cuticle water contents.

cell by means of different rectangular areas of filter. Individual samples too small to weigh accurately can be quantified by the same area-measuring technique after they are first weighed in bulk.

V. Future Developments

The fact that water vapor is absorbed by means of identifiable organ systems opens the possibility of studying them in the isolated state. The *in vitro* study of insect water-transporting organs—such as midgut, rectum, Malpighian tubules, and salivary glands—is already highly developed and successful. In at least some instances essentially the same water-transporting organs have become modified to transport water vapor. It seems possible, therefore, that the wide range of techniques now used in epithelial transport (e.g., tracers, drugs, and microelectrodes) might also be applied in vapor-transporting tissues. Those workers who have begun applying these techniques to vapor-absorbing organs have avoided the technical problem of delivering regulated humidifed air. Electrical contact with the rectal wall in *Thermobia* could only be achieved by filling the rectum with 0.5*M* KCl (Küppers and Thurm 1980). In a somewhat more realistic representation of vapor transport, Machin (1982) filled the rectal lumen of *Tenebrio* with electrolyte-free sucrose solutions. In some systems it may be possible to supply the isolated organ with water vapor using the method developed by Rudolph and Knülle (1978), in which the volume of a drop of saturated salt solution is monitored to measure vapor flux.

The importance of *in vitro* techniques is that more physiological information can be obtained than is possible in the intact animal and the limits of the uptake process more readily explored. More complex physiological information leads to more accurate mathematical modeling of the complex dynamic properties of the vapor uptake mechanism. A fuller understanding of vapor uptake kinetics may also shed light on the final phases of the uptake process, i.e., water absorption into the blood. Much more is known at present about the initial phase of vapor uptake—condensation—because it is more accessible to experimentation.

References

Arlian LG (1975) Dehydration and survival of the European house dust mite, *Dermatophagoides pteronyssinus*. J Med Entomol **12**:437–442

Arlian LG, Wharton GW (1974) Kinetics of active and passive components of water exchange between the air and a mite, *Dermatophagoides farinae*. J Insect Physiol **20**:1063–1077

Beament JWL, Noble-Nesbitt J, Watson JAL (1964) The water-proofing mechanism of arthropods. III. Cuticular permeability in the firebrat, *Thermobia domestica* (Packard). J Exp Biol **41**:323–330

Campbell EC, Campbell GS, Barlow WK (1973) A dewpoint hygrometer for water potential measurement. Agric Meterol **12**:113–121

Dessens H (1946) La brume et brouillard étudies à l'aide des fils d'araignées. Ann Geophys **2**:276–278

Devine, TL, Wharton GW (1973) Kinetics of water exchange between a mite, *Laelaps echidnina,* and the surrounding air. J Insect Physiol **19**:243–254

Dunbar BS, Winston PW (1975) The site of active uptake of atmospheric water in larvae of *Tenebrio molitor.* J Insect Physiol **21**:495–500

Edney EB (1971) Some aspects of water balance in Tenebrionid beetles and a thysanuran from the Namid Desert of Southern Africa. Physiol Zool **44**: 61–76

Edney EB (1977) Water balance in land arthropods. Springer, New York

Gilby AR (1980) Transpiration, temperature and lipids in insect cuticle. Adv Insect Physiol **15**:1–33

Hew CL, Slaughter D, Fletcher GL, Shashikant BJ (1981) Antifreeze glycoproteins in the plasma of Newfoundland Atlantic cod (*Gadus morhua*). Can J Zool **59**:2186–2192

Johansson B (1920) Der Gaswechsel bei *Tenebrio molitor* in seiner Abhängigkeit von der Nahrung. Acta Univ Lund **16**:1–36

Knülle W, Devine TL (1972) Evidence for active and passive components of sorption of atmospheric water vapor by larvae of the tick *Dermacentor variabilis.* J Insect Physiol **18**:1653–1664

Knülle W, Spadafora RR (1970) Occurrence of water vapor sorption from the atmosphere in larvae of some stored product beetles. J Econ Entomol **4**: 1069–1070

Krogh A, Weis-Fogh T (1951) The respiratory exchange of the desert locust (*Schistocerca gregaria*) before, during and after flight. J Exp Biol **28**:344–358

Küppers J, Thurm U (1980) Water transport by electroosmosis. In: Locke M, Smith DS (eds) Insect biology in the future. Academic Press, New York

List RJ (1958) *Smithsonian Meterological Tables,* 6th rev. ed. Smithsonian Institution, Washington, D.C.

Leighly J (1937) A note on evaporation. Ecology **18**:180–198

Loveridge JP (1980) Cuticle water relations techniques. In: Miller TA (ed) Cuticle techniques in arthropods. Springer, New York

Ludwig D, Anderson JM (1942) The effects of different humidities at various temperatures on the development of four moths. Ecology **23**:259–274

Machin J (1975) Water balance in *Tenebrio molitor,* L. larvae; The effect of atmospheric water absorption. J Comp Physiol **101**:121–132

Machin J (1976) Passive exchanges during water vapor absorption in mealworms (*Tenebrio molitor*): A new approach to studying the phenomenon. J Exp Biol **65**:603–615

Machin J (1978) Water vapor uptake by *Tenebrio:* A new approach to studying the phenomenon. In: Bolis L, Schmidt-Nielsen K, Maddrell SHP (eds) Comparative physiology: Water, ions and fluid mechanics. Cambridge University Press, Cambridge, England

Machin J (1979a) Compartmental osmotic pressures in the rectal complex of

Tenebrio larvae: Evidence for a single tubular pumping site. J Exp Biol **82**: 123–137

Machin J (1979b) Atmospheric water absorption in arthropods. Adv Insect Physiol **14**:1–48

Machin J (1982) Water vapor absorption in insects. Am J Physiol 244 (Regulatory Integrative Comp Physiol 13):R187–R192.

Machin J, O'Donnell MJ, Coutchie PA (1982) Mechanisms of water vapor absorption in insects. J Exp Zool **222**:309–320.

Mason EA, Monchick L (1965) Survey of the equation of state and transport properties of moist gases. In: Wexler A (ed) Humidity and moisture, Vol. 3. Reinhold, New York

Noble-Nesbitt J (1969) Water balance in the firebrat, *Thermobia domestica* (Packard). Exchanges of water with the atmosphere. J Exp Biol **50**:745–769

Noble-Nesbitt J (1970a) Water uptake from subsaturated atmospheres: Its site in insects. Nature **225**:753–754

Noble-Nesbitt J (1970b) Water balance in the firebrat, *Thermobia domestica* (Packard). The site of uptake of water from the atmosphere. J Exp Biol **52**:193–200

Noble-Nesbitt J (1975) Reversible arrest of uptake of water from subsaturated atmospheres by the firebrat, *Thermobia domestica* (Packard). J Exp Biol **62**:657–669

Noble-Nesbitt J (1978) Absorption of water vapor by *Thermobia domestica* and other insects. In: Bolis L, Schmidt-Nielsen K, Maddrell SHP (eds) Comparative physiology: Water, ions and fluid mechanics. Cambridge University Press, Cambridge, England

O'Donnell MJ (1977) Site of water vapor absorption in the desert cockroach, *Arenivaga investigata.* Proc Natl Acad Sci USA **74**:1757–1760

O'Donnell MJ (1980) Water vapor absorption by the desert burrowing cockroach *Arenivaga investigata* (Dictyoptera: Polyphagidae). Doctoral dissertation, University of Toronto

O'Donnell MJ (1981a) Frontal bodies: Novel structures involved in water vapor absorption in the desert burrowing cockroach, *Arenivaga investigata.* Tissue Cell **13**:541–555

O'Donnell MJ (1981b) Fluid movements during water vapor absorption by the desert burrowing cockroach, *Arenivaga investigata.* J Insect Physiol **27**: 877–887

O'Donnell MJ (1981c) Water vapor absorption by the desert burrowing cockroach, *Arenivaga investigata:* Evidence against a solute-dependent mechanism. J Exp Biol **96**:251–262

O'Donnell MJ (1982) Hydrophilic cuticle–the basis for water vapour absorption by the desert burrowing cockroach, *Arenivara investigata.* J Exp Biol **99**:43–60

Okasha AYK (1971) Water relations in an insect, *Thermobia domestica.* II. Relationships between water content, water uptake from subsaturated atmospheres and water loss. J Exp Biol **57**:285–296

Ramsay JA (1964) The rectal complex of the mealworm *Tenebrio molitor* L. (Coleoptera, Tenebrionidae). Philos Trans Soc Land (Biol) **248**:279–314

Richards JM (1971) Simple expression for the saturation vapour pressure of water in the range $-50°$ to $140°$. Brit J Appl Phys **4**:L15–L18

Rudolph D (1982a) Occurrence, properties and biological implications of the active uptake of water vapour from the atmosphere in Psocoptera. J Insect Physiol **28**:111–121

Rudolph D (1982b) Site, process and mechanism of active uptake of water vapour from the atmosphere in the Psocoptera. J Insect Physiol **28**:205–212

Rudolph D, Knülle W (1974) Site and mechanism of water vapor uptake from the atmosphere in ixodid ticks. Nature **249**:84–85

Rudolph D, Knülle W (1978) Uptake of water vapor from the air: Process, site and mechanism in ticks. In: Bolis L, Schmidt-Nielsen K, Maddrell SHP (eds) Comparative physiology: Water, ions and fluid mechanics. Cambridge University Press, Cambridge, England

Rudolph D, Knülle W (1982) Novel uptake systems for atmospheric water vapor among insects. J Exp Zool **222**:321–333

Schwertz PA, Brow JE (1951) Diffusivity of water vapor in some common gases. J Chem Phys **19**:640–646

Solomon AK (1960) Compartmental methods of kinetic analysis. In: Comar CL, Bronner F (eds) Mineral metabolism. Academic Press, London

Solomon ME (1951) The control of humidity with KOH, H_2SO_4, and other solutions. Bull Entomol Res **42**:543–559

Weatherley PE (1960) A new micro-osmometer. J Exp Bot **2**:258–268

Wharton GW, Arlian LG (1972) Utilisation of water by terrestrial mites and insects. In: Rodriguez JG (ed) Insect and mite nutrition. North-Holland, Amsterdam

Wharton GW, Devine TL (1968) Exchange of water between a mite, *Laelaps echidnina,* and the surrounding air under equilibrium conditions. J Insect Physiol **14**:1303–1318

Winston PW, Bates DH (1960) Saturated solutions for the control of humidity in biological research. Ecology **41**:232–237

Chapter 5
Microrespirometry in Small Tissues and Organs

K. Sláma

I. Introduction

Endothermic biological oxidations that are terminated by environmental oxygen are by far the predominate sources of energy in tissues or cells of insects; therefore, any more extensive metabolic conversion in the body should affect the intensity of respiration in one or another tissue. Respirometric data are less specific than the usual biochemical criteria for measuring intermediary cellular metabolism: they reflect the summation of all metabolic processes operative in the complex system of tissues and organs. It is generally accepted, however, that the changes found in tissue respiration are the best indication of metabolic changes associated with growth, development, and the maintenance or performance of physiological or biochemical functions. Although the literature contains extensive data on respiration of whole insects (for review see Keister and Buck 1974), our knowledge concerning rates of tissue respiration is incomplete.

The basic principle of the respiratory process depends on a physiologically controlled generation of free protons from endogenous substrate by a complex of dehydrogenating enzymes. The liberated electrons are then carried through a chain of the cytochrome enzymes to oxygen atoms. Metabolic water is formed and the resulting endothermic energy is stored in the form of high-energy phosphate bonds. Within any cell the whole respiratory apparatus resides on mitochondrial membranes. Isolated mitochondria are capable of performing extensive respiratory functions when provided with some exogenous substrate (for review see Sacktor 1974). Biochemical studies on isolated mitochondria have revealed valuable information on the nature and properties of the respiratory enzymes involved and the substrates being utilized. These data have, however, never been successfully used for the estimation of physiological rates of cell respiration.

In most experimental setups, the tissue selected for respirometry has to be removed from the hemolymph and disconnected from the regular supply of oxygen involving the tracheal system. Such explanted tissue will definitely exhibit some kind of respiratory activity even when kept in pure physiological saline. However, the extent to which this "explanted" respiration is related to that occurring in the intact body depends on a complex of intrinsic and extrinsic factors. It may depend, for instance, on the ionic strength and composition of the incubation medium, the presence of the proton-donating metabolic substrate, the availability of oxygen to the mitochondria, and the respirometric procedures employed.

Some tissue containing large reserves of glycogen or fatty acid (muscle, fat body) may exhibit relatively constant respirometric values for several hours after explantation. Other tissues containing smaller reserves of endogenous substrate (intestinal epithelium, salivary glands, Malpighian tubules, pericardial cells, endocrine glands) usually decrease their respiration within a few minutes after explantation into physiological saline. In the latter case respiration can be enhanced by the addition of an exogenous substrate that is generally known to increase respiration of mitochondria (Krebs citric acid cycle intermediates, amino acids, hexoses, α-glycerophosphate, synthetic dyes). Since different tissues show different substrate specificity, the data obtained in this manner are useful only for determining relative changes in the respiration of some particular tissues.

In addition to the above-mentioned problems with substrate specificity, further obstacles in respirometry of insect tissues and organs are associated with the limited diffusion of oxygen to mitochondria. The main point here is that the diffusion of oxygen through even the most permeable of tissues is only about 1/1,000,000 as fast as in the gas phase (Buck 1962). Some tissues with very high respiratory rates (muscle, nerve ganglia) show a rich supply of the tracheoles that terminate within 10–40 μm from the mitochondria (see reviews by Buck 1962; Miller 1974). Since most insect epithelia consume less oxygen than a contracting muscle, a satisfactory distance for the diffusion of oxygen to cover their metabolic demands may be about 50–100 μm. Provided that this estimation is correct, it would appear to be quite unrealistic to attempt to obtain reliable respirometric data from the measurements on explanted organs of any larger size (i.e., smallest diameter greater than 0.2 mm).

The solubility of oxygen in various incubation media does not differ considerably from its solubility in water or tissues. The diffusion of oxygen through the medium is equally dependent on the exposed surface area and on the distances to be traversed through the liquid (see review by Mill 1974). In most of the manometric methods, oxygen diffusion through the liquid has been enhanced by shaking or stirring the medium (Umbreit et al. 1972). It appears, however, that the problem of oxygen diffusing to the mitochondria can be reduced by miniaturization of the tissue samples, so that oxygen passes through distances similar to those in the living body. It is obvious that this method requires the use of very sensitive respirometric techniques.

A brief analysis of the techniques used in microrespirometry is presented in this chapter. Two methods for measuring the respiration of small tissue and organ samples are described: one is a very simple, direct volumetric method that can be practiced by university students; the other is a more professional method that is recommended chiefly for tissue culture research.

II. Respirometric Techniques

Almost all of the general respirometric methods have been used at one time or another on insect material. Many of these techniques have been improved and miniaturized to accommodate the small size of the insect body. Although the techniques to be described were mainly employed to measure respiration of the whole body, by adaptation of the respirometers to small oxygen or CO_2 volumes they may be used for respirometry of small tissues as well.

1. Manometric Methods

The direct and indirect Warburg method (Fig. 5.1a) has had the widest application in respirometry of whole insects, insect homogenates, and explanted tissues. For detailed descriptions of the method see Dixon 1951; Glick 1949; Kleinzeller 1965; Umbreit et al. 1972. Advantages of this method are standardized equipment, the ability to run whole series of measurements simultaneously and the ability to operate with gas mixtures of different composition. Disadvantages are the necessity for exact calibration of the vessel volume, the influence of barometric pressure, and sensitivity limited to the range of microliters per reading.

Increased sensitivity has been obtained by modification of the Warburg system into differential manometers commonly known as Barcroft respirometers (Fig. 5.1b). These are in principle two respirometric compartments closed off from atmospheric pressure and connected by a manometer. The position of the manometric fluid indicates volumetric changes at any given moment. This principle allows miniaturization, although exact calibration of the vessel volume is still necessary for computation of the respirometric data.

A differential manometer for insect respiration was described by Crisp and Thorpe (1974). Several other microrespirometers (Cunningham-Kirk or Tobias-Gerard differential respirometers) based on the same principle were devised for respirometry on small tissues and organs (described in reviews by Glick 1949; Tobias 1943). The most sensitive of all manometric techniques (less than 1 nl per reading) is the Cartesian diver manometry (for detailed description see Glick 1949). It is suitable for work with single or just a few cells or with very small tissue fragments. A disadvantage is the professional skill required to cope with handling or filling the miniature divers.

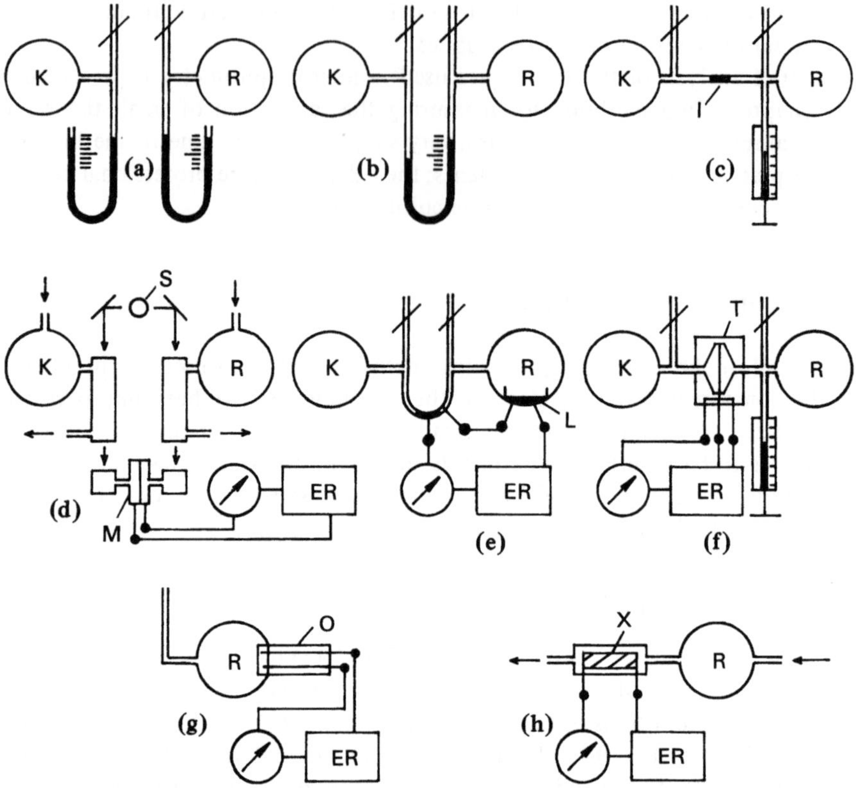

Fig. 5.1. Common principles used in insect respirometry. **(a)** Warburg manometers. K, compensatory vessel; R, respiratory vessel. **(b)** Differential manometer of the Barcroft type. **(c)** Compensatory volumetric principle. I, index solution. **(d)** Infrared gas analyzer for CO_2 release (Hamilton 1964); S, infrared radiator; M, differential temperature-sensitive detector; ER, electrical circuitry and recorder. **(e)** Electrolytic respirometer. L, $CuSO_4$ solution for electrolytic production of oxygen. **(f)** Continuously recording volumetric respirograph. T, differential semiconductor pressure transducer; the syringe is used only for calibration. **(g)** Polarographic respirometer. O, oxygen electrode. **(h)** Diaferometric or gas chromatographic flow-through respirometer. X, thermal conductivity detector.

2. Volumetric Methods

The direct volumetric principle of respirometry is derived from differential manometric methods. Scholander (1942, 1950) adapted volumetric methods to insect material (see also Glick 1949; Scholander et al. 1951). According to Scholander and Iversen (1958) there are certain merits inherent in volumetric microrespirometry that other systems lack, namely, calibration is independent of vessel volume and the system can easily compensate for small temperature and pressure variations. Scholander (1950) described the common volumetric principle as follows: "The respiration chamber with CO_2 absorber is balanced against a compensating chamber through a manometer. As oxygen is being used it is replaced from a measuring device so as to maintain balance with the compensating chamber. The oxygen consumed is read directly on the measuring device" (Fig. 5.1c).

Numerous types of volumetric respirometers for work with insects have been described. The design was improved by replacing the whole body of the instrument with plastic, by the invention of micrometric burettes, syringes, or automatic volumetric burettes, and by the use of shaking racks with multiple respirometers (Gilson 1963; Scholander and Iversen 1958; Scholander et al. 1951). The sensitivity of all the volumetric microrespirometers is mainly determined by the capacity of the respiratory vessels. The decrease of this capacity to 100 μl allows direct measurements of oxygen consumption at nanoliter ranges. The recent availability of very thin plastic tubing, miniature valves and joints, and precision microsyringes (Hamilton Corp., Reno, Nev.) makes the construction of the instruments easy and makes the technique of volumetric microrespirometry ideally suited for measurements on small tissues and organs.

3. Optical Methods

Hamilton (1959, 1964) applied infrared gas analysis techniques to the measurement of respiratory CO_2 output in insects (Fig. 5.1d). An air stream is passed first through a respiratory tube containing the insect and then through an infrared gas analyzer. Changes in CO_2 concentration (from 0 to 0.02%) cause corresponding changes in the absorbance of the infrared radiation. This affects the balance of a thermal detector and signals from the detector are recorded. The sensitivity of the instrument is in the range of microliters per minute. The sensitivity of the infrared gas analysis method has been compared with that of some other respirometric techniques by Wightman (1977).

A miniature differential manometer operating on an optical principle was described by Tobias (1942). The liquid index is replaced by a thin glass membrane coated with a film of collodion. When illuminated with monochromatic light, displacement of the membrane produced by changes in internal pressure cause shifts of interference fringes and their magnitude is recorded. The instrument is sensitive to pressure changes of 5 mPa.

4. Electrical Methods

Due to their great sensitivity, polarographic methods have been commonly used to measure dissolved oxygen in biological samples (Umbreit et al. 1972). Depending on the construction of the measuring cells and electrical properties, they can reliably determine changes in concentration that are equivalent to nanoliter volumes of oxygen (see review by Degn et al. 1980).

The recent availability of commercially manufactured oxygen electrodes (Fig. 5.1g) with the measuring cell separated from the outside by oxygen-permeable polyethylene or Teflon membranes (Beckman Model 777 oxygen analyzer, Beckman Instruments, Fullerton, Calif.; also Yellow Spring Instruments Co.), along with thermistor compensation of temperature and high-quality electronics, make this method universal for oxygen measurements in biochemical analysis (see review by Lessler and Brierley 1969). The method has been used to measure respiration of whole insects (Gilby and Rumbo 1980) or respiration of explanted fat body (Brown and Chippendale 1977). An advantage is a small dependence on changes of mechanical pressure, and a disadvantage is that the readings are obtained in terms of oxygen concentration and the volume must be computed.

Electrolytic respirometers (Fig. 5.1e) are favorite instruments for prolonged measurements and continuous recording of oxygen consumption. They are mostly constructed on the principle of a constant-volume differential manometer. The movement of the manometric fluid acts electrically or optically on contacts that send electrical current to special electrodes placed in $CuSO_4$ solution inside the respiratory vessel. The current required to compensate the missing volume by electrolytically produced oxygen is then recorded. A good survey of the construction of the various electrolytic respirometers has been provided by Klekowski and Zajdel (1972). A continuously recording instrument that is sensitive to 70 nl O_2 h^{-1} and is suitable for work with small insects has been described by Fourche (1964) (see also Kuusik 1977; Taylor 1977; Winteringham 1959; Heusner and Tracy, this volume, Chapter 7).

Other electrical methods of respirometry include diaferometry and electrical transducers of mechanical pressure. The diaferometry uses the flow of air, which is passed through a tube containing an electrically heated platinum wire. Changes in the gas composition of the air stream produce resistance changes of the wire and this is recorded by a linear recorder (Fig. 5.1h). The method has the same advantages and disadvantages of flow-through systems. It has been successfully used for the demonstration of discontinuous CO_2 outbursts in insects (Punt 1950, 1956). Its sensitivity is too small to merit consideration for work with small tissue samples (Wightman 1977).

The possible use of absolute pressure transducers in respirometry was hindered mainly by their inconvenient design, large size, insufficient temperature compensation, and a generally poor signal-to-noise ratio at relative pressure changes smaller than 10 Pa. Gilby and Rumbo (1980) recently used an absolute pressure transducer (LX 1600 A, National Semiconductors Co., USA) to measure oxygen consumption in the fly puparia. Because of a low signal-to-noise ratio, it was used only as a supplementary device to an oxygen electrode. We have recently developed a sensitive differential semiconductor transducer, especially designed for respirometric studies. It has been used for the construction of the microrespirograph (Fig. 5.1f) described in detail in Sect. VIII.

5. Chemical and Radiometric Methods

Thermal conductivity detectors of great sensitivity used in gas chromatography have also found an application in insect respirometry (Putman 1976; Tadmor et al. 1971). The gas chromatography method has certain advantages over other flow-through techniques (diaferometry, infrared gas analysis), but its sensitivity is greatly inferior in comparison to the volumetric constant-volume systems (Kuusik 1976; Wightman 1977). By contrast, the radiometric techniques, which determine the rate of $^{14}CO_2$ released by tissues incubated with the radiolabeled exogenous substrate, are extremely sensitive (see review by Wang 1976). Unfortunately, these results merely give information on the relative changes in metabolism of specific substrates (Wiens and Gilbert 1965).

III. Essential Features of Tissue Respiration

The comparative oxygen consumption data for various species and developmental stages given by Keister and Buck (1974) indicate that (except for very low values in diapausing stages and extremely high values in flying insects) most of the oxygen consumption rates fall within the limits of 100–1000 $\mu l\ h^{-1}\ g^{-1}$ live weight. These amounts of oxygen are almost exclusively used and converted to metabolic water in mitochondria inside the cells.

The insect body also has more or less extended intra- and extracellular spaces containing no mitochondria (hemolymph, lumen of the gut, sclerotized cuticle, accumulated reserve materials in some tissues) that contribute little or nothing to active respiration. It is likely that tissues with well-developed mitochondrial systems would exhibit somewhat higher rates of respiration than the whole body (muscle, nerve, epithelia), in contrast to some storage organs (salivary glands, spinning glands, ovaries) which contain large amounts of the metabolically inert material. In addition, there are striking changes in respiratory metabolism of a given tissue when investigated under different physiological conditions. However, in spite of such variability, the majority of oxygen consumption rates of the explanted tissues measured thus far (see Table 5.1 below) have also occurred within the above-indicated limits for respiration of the whole body. Thus, to estimate reasonably the required sensitivity of the respirometer to be used, we can calculate that during each hour an explanted insect tissue would most likely consume the amount of oxygen equivalent to 0.1-1.0 of its volume. In other words, 1 mg (μl) of tissue would likely consume about 0.1-1.0 $\mu l\ O_2\ h^{-1}$.

1. Respiration of Homogenates

Respiration of tissue homogenates has been used occasionally as a crude measure of tissue metabolism. Bodine (1950) and his co-workers were the first to study to what extent the oxygen consumption of the whole body was related to that

Table 5.1. Rates of Oxygen Consumption in Explanted Tissues and Organs of Some Representative Insects

Species	Tissue	Stage	Method[a]	Medium[b]	O_2 consumption (nl h^{-1} mg^{-1} live weight)[c]	Reference
Leucophaea maderae	Thoracic muscle	Adult	W	R	650–670	Samuels 1956
	Prothoracic glands	Larva	VR	R	105	Bernardini and Laudani 1966
	Fat body	Adult	VR	R	200–400	Müller and Engelmann 1968
Nauphoeta cinerea	Fat body	Adult	W	R	500	Lüscher 1968
Blaberus discoidalis	Fat body	Adult	W	R	235	Keeley and Friedman 1967
	Muscle	Adult	W	R	1680	Keeley and Friedman 1967
Diatraea grandiosella	Fat body	Larva	OE	RA	100–125	Brown and Chippendale 1977
Bombyx mori	Ovaries	Pupa	W	R	250–620	Fourche and Ambrosioni 1969
Drosophila hydei	Salivary glands	Larva	W	R	157–280	Leenders and Knoopien 1973
Pyrrhocoris apterus	Corpus allatum	Adult	VR	R	1050	Sláma (unpublished)
			VR	GM	2450	

[a]W, Warburg method; VR, volumetric respirometer; OE, oxygen electrode.
[b]R, Ringer's solution; RA, Ringer's solution with substrates; GM, Grace's tissue culture medium.
[c]Values based on dry matter converted to live weight by 1:4 ratio.

of its homogenate. They used grasshopper embryos homogenized by glass homogenizers in phosphate-buffered Ringer's solution. The oxygen determination was carried out by the standard Warburg technique. It appeared that respiration of the homogenate was more or less proportional to that of the intact embryo (Bodine 1950), but its intensity was decreased by 65%. Further studies revealed that different intracellular constituents respired with different rates (Bodine and Lu, 1950) and the respiratory quotient was increased in the homogenate from 0.7 to 1.0 (Bodine and West 1953).

Bodine's studies were extended to other insects by Ludwig and Barsa (1956) using similar Warburg techniques. They confirmed the reduction of respiration of homogenates versus that of the whole body and noticed that the reduction was strongly dependent on the dilution of the homogenate (from 29% reduction with 20% homogenate to 64% reduction with 1% homogenate). Addition of succinate and cytochrome *c* restored the respiration of the homogenate almost to the level of the intact body (Ludwig and Barsa 1957). Since that time, considerable knowledge of the effects of added substrates and inhibitors on homogenate or mitochondrial respiration has been gained (see review by Sacktor 1974). It appears that homogenates prepared from different tissues show different substrate dependence, so that work with homogenates cannot be standardized for all tissues. Because results obtained from measurements of homogenate respiration give only approximations of the respiration of intact or explanted tissues, their physiological value is questionable.

2. Explanted Tissues and Organs

Isolated fragments of the insect body continue to respire for many hours or days when prevented from desiccation (Sláma 1965). The tissues and organs of the fragment still receive a regular supply of oxygen through the tracheal system and a regular supply of metabolic substrate from the hemolymph. These basic physiological conditions are severely disturbed whenever a tissue is removed from the body and transferred into a liquid medium. It is obvious that the source of metabolic substrate, the ionic balance across the cytoplasmic membrane, the redox potential balance, and the buffering capacity of the hemolymph may all be lost by selection of an inappropriate incubation medium. Glycogen, which is used as a universal fuel for energy metabolism, slowly leaks out of the explanted cells. The tracheal system of small tissues is invariably blocked by fluid during explantation. Oxygen has to diffuse from outside the cells, at first through an extracellular stromal sheath of varying thickness and composition, and then through the cell membrane, before reaching the mitochondria. Therefore, two basic experimental conditions must be met to get any reasonable physiological data on tissue respiration: (1) satisfactory diffusion of oxygen to the mitochondria, and (2) adequate substitution of the hemolymph properties by the incubation medium.

The most commonly employed Warburg method facilitates oxygen diffusion so that the tissue is freely floating in the medium, which is continuously mixed by shaking with the manometers. Vigorous shaking may cause damage to some fragile tissues. Occasionally, it was not the respiration of living cells that had been measured, but that of the cell remnants similar to homogenate. As mentioned above, an alternative to shaking is to shorten the paths of oxygen diffusion through the liquid to the distances observed in the living body. This is achieved by using microliter instead of milliliter amounts of the medium and small tissue samples whose thickness does not exceed 200 μm. These conditions can be easily fulfilled with most insect epithelial tissues.

Insect hemolymph cannot be used as a respirometric medium without special precautions. In particular, activation of phenoloxidase enzymes can result in spontaneous oxygen consumption by the hemolymph. When placed into a respirometer, 1 μl of insect hemolymph would usually consume more oxygen than an equivalent amount of tissue. This self-oxidation ability of the hemolymph can be reduced to some extent in the respirometer by addition of phenylthiourea or some other inhibitor of phenoloxidase. In this case, the possibility of causing adverse effects on tissue respiration cannot be excluded. Under certain circumstances a clear supernatant obtained by centrifugation of the heat-coagulated hemolymph can be stored in a refrigerator for later use as a respirometric medium. According to my observations, however, the samples so prepared from the hemolymph of one developmental stage cannot be successfully used as a universal medium for tissues explanted from other stages or species.

Buffered insect Ringer's solution has often been used as a respirometric medium under the assumption that the explanted tissue might contain enough endogenous substrate to support respiratory metabolism. In most cases these studies employed the standard Warburg technique with relatively large amounts of tissue, excess of Ringer's solution, and vigorous shaking at 25°–30°C. Some of these results are summarized in Table 5.1. Due to large differences in assay conditions it is difficult to assess the physiological significance of these data. It is reasonable to assume that *in vivo* rates of tissue respiration would be somewhat higher. Nevertheless, the majority of the data in Table 5.1 do not deviate from the common range of respiration intensity of the whole insect body mentioned above, i.e., 100–1000 $\mu l\ O_2\ g^{-1}\ h^{-1}$ (100–1000 $nl\ O_2\ mg^{-1}\ h^{-1}$).

A decrease in oxygen consumption due to depletion of endogenous substrate has been less pronounced in explanted muscle (Samuels 1956) or fat body (Lüscher 1968; Müller and Engelmann 1968), but more pronounced in explanted glandular tissues. Thus, for instance, respiration of the prothoracic glands was increased 5–7 times by the addition of glucose to Ringer's solution (Bernardini and Laudani 1966), or that of explanted salivary glands was increased to about 200% by the addition of certain amino acids or intermediates of the citric acid cycle (Leenders and Knoopien 1973).

Using the microvolumetric methods described in Sect. VII and VIII below, I have investigated the effect of various incubation media on respiration of very small insect organs (corpora allata, 10 nl volume). After explantation into

Ringer's solution, and 15-20 min for temperature equilibration, the oxygen consumption of the organs always diminished over a period of 30-60 min to about one-fourth of the first reading and then remained more or less constant for several hours. Although the initial decrease could be partly or completely eliminated by the addition of exogenous substrates of mitochondrial respiration (succinate, NADH + NADH-dependent substrates), the most satisfactory and reproducible results were obtained with several tissue culture media that are currently used for maintaining insect epithelia (for composition and review see Oberlander 1980).

In osmotically balanced Grace's medium explanted organs exhibited steady rates of respiration for several hours, with CO_2 values considerably higher than in the first readings in Ringer's solution (Table 5.1). The tissue culture media contain all the important substrate constituents (i.e., hexoses, citric acid cycle intermediates, amino acids, salts, and vitamins). The explanted tissues in these media not only survive, but in the presence of the hormones, they are able to perform normal developmental cycles and physiological functions (Oberlander 1980). Tissue culture media as standard respirometric media have further practical utility for respirometric control of tissues cultured *in vitro*.

IV. Bases for Expressing Rates of Tissue Respiration

The common bases for expressing the rates of oxygen consumption or CO_2 release are per milligram of protein, of dry weight, or of live weight. The unit of protein as a basis is very practical but it has a biochemical rather than physiological justification. Because of drastic changes in the protein transfer between insect organs, the respirometric data expressed per unit of protein may be misleading. Assuming, for instance, that a fat body cell has a constant respiration, then the rates expressed per milligram of protein would indicate very low values for cells with accumulated protein reserves and, conversely, very high values for the cells having depleted reserves. This general physiological problem obviously applies to a limited extent to the units of volume or weight as well.

The relationship can perhaps be best illustrated on the growing ovaries of *Bombyx* (Fourche and Ambrosioni 1969). Here oxygen consumption of the organ dramatically increases during ovarian growth, but the size of the organ increases even more rapidly due to deposition of yolk into the oocytes. The respiratory rate expressed per unit of weight paradoxically shows the highest values for the small inactive organ, decreasing values during the most intensive growth period, and lowest values for the large ovaries.

We have to realize that respirometric data are mostly useful for physiological studies. In this respect the best basis for their expression would be some anatomically or physiologically defined structure, such as a single cell, small organ, anatomically determined part of an organ, anatomically defined part of the tissue, etc. This information would give the best account of changes in respiration during development, but it is not sufficient for making comparisons between metabolic intensities of different tissues. Therefore, the unit of weight or volume cannot be avoided.

According to Keister and Buck (1974), the unit of live weight appears to be as good a basis as any for expressing insect respiratory rates in general. Manipulation with very small tissue samples unfortunately almost excludes the possibility of determining their weight by the usual gravimetric methods. The more practical solution is to calculate weight of very small tissue or individual cells indirectly by means of the microvolumetric method described in the next section.

V. Microvolumetric Determination of Size of Small Tissues

The microvolumetric method is based on the principle of calculating volume from the surface area occupied by a tissue layer of a constant height. The small pieces of tissue selected for respirometry are dissected in isotonic saline under the stereomicroscope. With a small amount of the saline they are transferred to the central part of hematological slides (Bürker chambers that are used for hemocyte counts), which are calibrated to give a constant distance between the cover slip and the bottom of the slide (10–100 μm). The required height of the layer is selected according to a preliminary inspection of the thickness of the epithelial wall or according to the smaller diameter of the organ or cell. The tissue must be slightly compressed when the cover slip is in place. The surface area occupied by the sample is projected onto a calibrated sheet of paper by means of a camera lucida or any type of projecting stereomicroscope. The area of the paper so covered is then excised and weighed. The volume (V in nanoliters) of the tissue or organ is then calculated using the equation: $V = k \cdot w$ where w is the weight of the excised paper in milligrams and k is a constant. The value of k for each experimental series can be calculated:

$$k = \frac{H}{w_1 \cdot m^2}$$

where H is the height of the tissue layer in micrometers, w_1 is the average weight of 1 mm² paper sheaths in milligrams and m is the magnification (for 1:120, $m = 120$).

The method is relatively simple and it can be performed routinely in a few minutes. The surface area on the paper can also be measured planimetrically. The difficulty is to obtain hematological slides of the height needed to accomodate the layer of tissue. The commercially available Bürker chambers are usually 100 μm, which is too large for epithelial tissues of small insects.

The desired height (usually 20–80 μm) can be obtained by gently abrading the side strips supporting the cover slip of Bürker chambers using fine carbide powder for glass polishing. The abrasion is inspected continuously by micrometer measurements. After abrasion, the correct height is calibrated by measuring the surface area occupied by a drop of mercury of known volume (1.36 mg = 100 nl). Repeated measurements with mercury have indicated less than a 2% standard error.

When a volume of tissue is measured after respirometry it is possible to stain the tissue (methylene blue, vital dyes) for better visualization of the surface area. With a layer 40 μm heigh I have routinely measured volumes of insect epithelia from 10 to 100 nl. By using a correction factor for tissue density, nanoliter volume data can easily be converted to micrograms of live weight.

This method has been adapted for microrespirometry of small tissue samples used to determine the tonicity of the respirometric incubation media. To this end the tissue is carefully dissected and its volume is measured in a drop of hemolymph. Then it is transferred into insect Ringer's solution (see Jones 1977 for selection of the solution) or the respirometric medium; its volume is again measured after 1 h of incubation. The change in volume indicates the hyper- or hypotonicity of the saline or medium.

VI. Preparation of Tissues; Conditions for Microrespirometry

A convenient method for immobilizing insects before dissection is to submerse them under water for 15-30 min. The use of narcotics (diethylether, chloroform, ethyl acetate) is prohibited due to their inhibitory effects on the mitochondrial enzymes. Narcosis by gaseous CO_2 should also be avoided whenever possible because prolonged exposures to high CO_2 concentrations disturb the buffering capacity of tissues, resulting occasionally in damage to some epithelial cells.

The immobilized insects are then dissected in saline at room temperature. The required portions of tissue or organs are washed from the hemolymph and transferred to slides for volume determination as indicated in Sect. V, or they are directly transferred to the incubation medium and placed in the respirometer. The transfer is made by constricted pipettes or by means of small loops made from a thin platinum wire. All of the solutions must be free of bacteria, whose presence in larger quantities may be responsible for erroneously high respirometric readings. When long-term measurements are to be made (over 5 h) the conditions for aseptic work common to tissue culture methods (Oberlander 1980) should be followed.

Selection of the size of the tissue sample to be measured depends on the respirometric method to be employed. The standard Warburg method requires milligram quantities of tissue in order to obtain microliter readings of oxygen consumption per hour. The procedures employed with the Warburg method have been described elsewhere (see Umbreit et al. 1972 for review and references in Table 5.1).

The microvolumetric methods described in Sect. VII are based on slightly different principles. They use microscopic amounts of tissue (10–200 nl) and minimum amounts of medium (0.5–2.0 μl). The medium is suspended in a small glass capillary to increase its surface. This arrangement allows oxygen to diffuse only through micrometer distances in the liquid phase so that a pO_2 equilibrium is established within a few minutes. The medium need not be stirred and, because of the small dimensions of the respiratory vessels (about 100–200 μl capacity), the CO_2 is rapidly absorbed by small amounts of low-concentration KOH. The thickness of the tissue samples that are freely floating in the medium should not exceed approximately 200 μm.

Additional conditions that have to be strictly observed in respirometry of small tissue samples are related to the absorption of expiratory CO_2. To facilitate CO_2 absorption, various authors have used very high concentrations of KOH (20% or more) or even solid KOH deposits on a filter paper. However, these conditions cannot be tolerated in a sensitive respirometer. The diffusion of CO_2 through the air is very rapid, and the absorption of CO_2 is chiefly dependent on the exposed surface of the KOH solution, and less on its concentration. Concentrated KOH solutions vigorously absorb water and thus decrease the water vapor tension in the respirometric flasks. As the KOH solution becomes diluted the tension of water vapor again increases, and these changes in water vapor tension may lead to erroneous oxygen consumption readings.

In respiratory vessels of smaller than 1-ml capacity, the use of concentrated KOH solutions can be completely avoided. The stoichiometric calculations reveal that only 1 μl 1% KOH can theoretically absorb up to 40 μl CO_2. This amount represents less than one-fifth of the total oxygen volume in a 1-ml respiratory vessel, which is beyond the capacity of a single respirometric experiment (note that the respiring object should not consume more than about 5–10% of the initial amount of oxygen present in the respiratory vessel). With respiratory vessels of 100-μl capacity I have determined that a 4-mm^2 area of 1% KOH would absorb 95% of CO_2 within 1.5 min at 25°C.

In the respirometry of small tissue samples the removal of water from the incubation medium by large concentrations of KOH would be especially problematic. The increased osmotic pressure of the medium would cause dehydration of the tissue sample. To avoid these unwanted effects the water vapor tension of the medium should be equilibrated with that of the KOH solution. This is achieved by measuring the transfer of water between these solutions within a closed compartment by volumetric, gravimetric, or osmometric methods.

VII. A Simple Volumetric Respirometer for Small Tissue

The respirometer shown in Fig. 5.2 is based on a common direct-reading volumetric principle (Sect. II.2). It can be easily constructed and used in almost any laboratory. Its sensitivity is approximately 2 nl per reading.

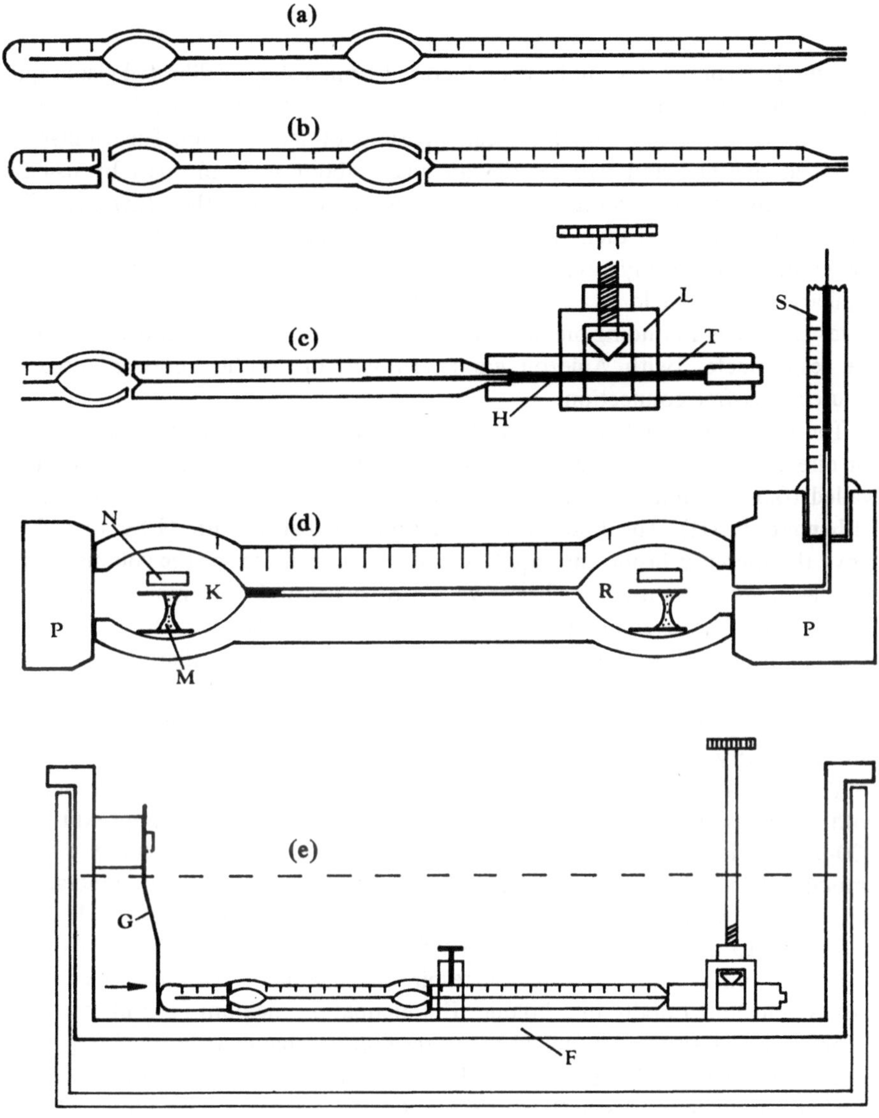

Fig. 5.2. Simple volumetric microrespirometer for measuring oxygen consumption of small tissue samples. **(a), (b)** Construction of the instrument from glass capillary. **(c)** Measuring device. T, silicone rubber tubing; H, mercury; L, mechanical pressure. **(d)** Detail of central part of respirometer with Hamilton 7001 N syringe as a measuring device. S, syringe; P, plastic blocks; K and R, compensatory and respiratory vessels; M, respiratory medium in a piece of glass capillary; N, porous plastic strip with KOH solution. **(e)** Whole respirometer in the circulating water bath. F, supporting rack; G, metallic spring holding parts together.

1. Description of the Instrument

The whole respirometer can be manufactured from a single calibrated glass capillary, such as that used for common thermometers. We used capillaries of 6–7 mm o.d., 0.08–0.1 mm i.d., with scale divisions of about 1 mm. The capillary is closed at one end and two cavities are made about 4–5 cm apart by using a narrow flame while simultaneously applying pressure from the other end (Fig. 5.2a). The capillary is then cut into three pieces as shown in Fig. 5.2b. The ends are ground and evenly polished on a rotating horizontal glass disk while the capillary is firmly held in a vertical position by a holder. Fine carbide powder in water is used as a grinding medium. The polished ends serve as joints to assemble the respirometer, with Apiezon grease, lanolin, or another lubricant used to ensure a leak-free connection.

The longer part of the capillary (Fig. 5.2c) is used as a compensatory device for direct readings of oxygen consumption. This is achieved by means of a thick-walled (0.5–1.0 mm i.d.) rubber or silicone rubber tubing (T), which is filled with mercury (H). The fine screw of the mechanical pressure (L) is turned to move the mercury inside the capillary and to compensate for the volume of the missing oxygen. We used capillaries with a capacity of 5–8 nl mm^{-1} length; for more accurate measurements it is possible to use capillaries up to 1 nl mm^{-1}. It seems more practical to replace the whole measuring capillary with a high-precision microliter syringe (Fig. 5.2d), such as a Hamilton 7001 N syringe with 1-μl total capacity (Hamilton Co., Reno, Nev.). Any other volumetric device with sensitivity in the nanoliter range can be used.

The central part of the respirometer (Fig. 5.2d) is a differential manometer with compensatory (K) and respiratory (R) chambers (6 mm i.d., approximately 100-μl capacity). The chambers are separated during measurements by a small amount of liquid (isooctane, isodecane, preboiled kerosene, light mineral oil, etc., stained with Sudan Black), which serves as a manometric index. It is administered to the orifice of the capillary from the compensatory side, where it is held by capillary force. Before the respiratory chambers are filled, the central part is held horizontally in a holder. The incubation medium (1–2 μl, need not be precisely measured) is supplied to the respiratory chambers in small fragments of glass capillaries (M, Fig. 5.2d), where it is held by capillary force. A small strip of porous inert plastic (Porvic S, 0.5 × 1.0 × 5.0 mm) soaked in KOH solution of equilibrated water vapor tension with the medium is stretched against the walls (N, Fig. 5.2d). The whole central part is then removed from the holder and assembled into the rack (Fig. 5.2e); care is taken not to apply warmth by touching the chambers with the fingers. The closed respirometer is held together by a spring (G). The racks with several respirometers are placed into the circulating water bath with temperature regulated to ±0.2°C.

The instrument is so simplified that it has no opening valves. Due to the limited capacity of the measuring device, it is best suited for small tissue samples of 30–200 nl volume (μg weight). When it is closed, care must be taken to avoid any larger temperature changes that would act only on one of the two compartments. Immersion into the water bath should therefore proceed suddenly and in a horizontal position. The tight joint connection can be checked before immersion by touching each compartment gently with a finger. The index should move slightly in the opposite direction.

When both the chambers are of equal capacity, the index will move about one-half the distance of the mercury in the measuring capillary. This is the limiting factor of the reading accuracy. The movement of the index and the accuracy can be increased when the compensatory chamber has larger capacity than the respiratory one. The reading accuracy is also highly improved when readings are made under a stereomicroscope. Due to the direct-reading volumetric principle, the central parts of several respirometers can be freely exchanged.

2. Working Procedure

A. Place a small drop of the index solution into the orifice of the capillary in a compensatory vessel (Fig. 5.2d).
B. Dissect the required portion of the tissue and measure its volume as described in Sects. V and VI.
C. Transfer the tissue into the respiratory medium and position it within the respiratory compartment. Place an equivalent amount of the medium without tissue into the compensatory compartment and supply the strips of porous plastic with isotonic KOH to both the compartments.
D. Adjust mercury in the measuring capillary to about one-quarter of its length (piston of the microsyringe to 0.75 μl).
E. Assemble the respirometer in the supporting rack and move the index slightly toward the center by manipulating with mechanical pressure (piston of the syringe).
F. Immerse the respirometer quickly into the water bath and allow temperature equilibration for 15 min.
G. Record the position of the index, meniscus of the mercury, or the position of the piston at a zero time (the index should slowly move towards the respiratory chamber when oxygen is being consumed).
H. Return the index to its original position in intervals of 10–60 min. The volumetric changes in the measuring device indicate nanoliters of oxygen consumed during the given period of time. Perform at least three successive readings.
I. Remove respirometer from water bath, blot the water out with absorbent paper, dismantle the parts, and repeat steps B–H with another tissue sample.

VIII. Universal Scanning Microrespirograph

The universal scanning microrespirograph, an electronic, continuously recording instrument, is built on a simple volumetric principle. It can be used for studying respiration kinetics in large animals, very small insects, insect eggs, isolated body fragments, or small tissue samples cultured *in vitro*. The manometric fluid between the respiratory and compensatory compartments is replaced here by a thin membrane bearing very sensitive semiconductor strain or tensometric sensors. The sensors transform changes of mechanical pressure into electrical signals that are amplified further and recorded. The measuring device that was used in normal volumetric respirometers for direct readings of oxygen consumption is here used only occasionally for calibration of the instrument.

1. Description of the Microrespirograph

The instrument consists of three major parts: the volumetric unit with the differential transducer, opening valve, and respiratory vessels; the electronic tensometric unit designed to transform changes in mechanical quantities to electrical signals; and the linear recorder (oscilloscope or any other electrical measuring device). The most essential part for respirometry is the volumetric unit with the transducer. This is described in detail, while the other parts, which are commercially available, are described only with the necessary technical data.

A. Volumetric Unit

The construction of the volumetric unit is shown in Fig. 5.3. There are two boxes (A and B) made from 3 mm brass plates.One houses the transducer (T) and the opening valve (V). The motor of the electrical valve driver (VD) is outside the box. The second box (B) accommodates the respiratory vessels (C and R). It can be opened from one side for exchange of the vessels.

The construction of respiratory vessels is optional. They can be drilled in one common block of Plexiglas with the cover lid bearing the outlets (L, Fig. 5.3); or they may be separate plastic test tubes, ordinary Warburg vessels, or special tissue culture tubes. Additional outlets (O) can be occasionally used for replacement of the inside air during prolonged measurements, or they can be used for calibration of the instrument (if the calibration syringe is not permanently attached to the system). The outlets of the transducer and the respiratory vessels are made of stainless steel needle material, 0.5 mm o.d. They are interconnected by Teflon tubing of 0.5 mm i.d., 1 mm o.d., with a capacity of about 1 μl 5 mm^{-1} length.

The miniature inert valve 1XP (Hamilton Co., Reno, Nev.) has one central and four side Luer-lock Teflon terminals. The central outlet is connected with the two opposite side outlets (V, Fig. 5.3) so that it opens and closes the respirometer with each 90° rotation. The valve is settled on the top of the electrical valve driver (Hamilton Co.) and the position of the valve is remotely controlled by a selector switch. In addition, the valve driver will rotate the valve for 360° (thus opening the respirometer two times) upon receiving the signal for full capacity of the scale reached on the recorder.

The construction of the differential pressure transducer is shown in Fig. 5.4. Its most essential component is the central plate (A, Fig. 5.4). It is made from a 1-mm metal-coated plastic plate used for manufacture of the printed circuits.

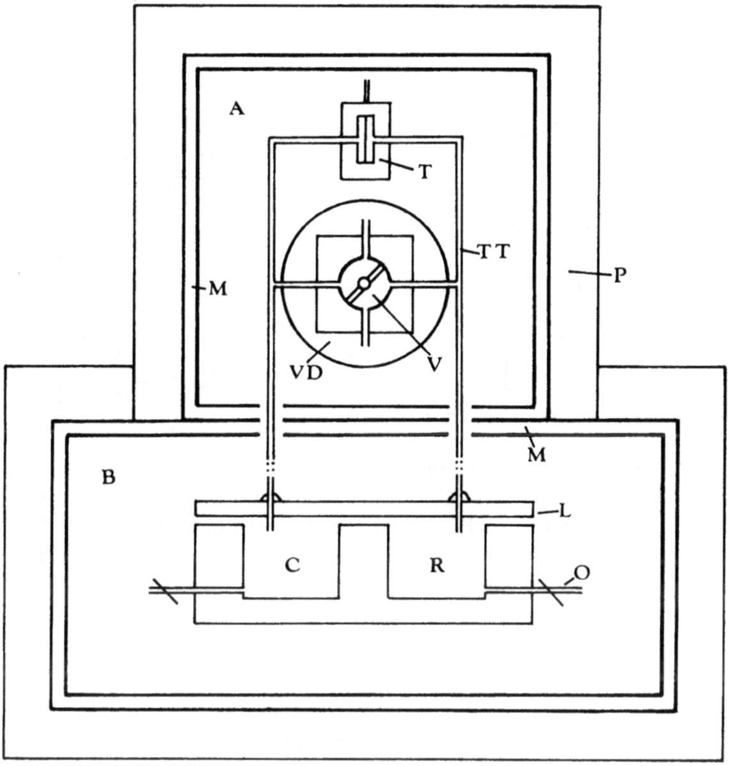

Fig. 5.3. Volumetric unit of the scanning microrespirograph for measuring respiration of small tissue samples. A, transducer and valve compartment; B, respiratory vessel compartment; T, differential pressure transducer; V, Hamilton 1XP valve with one central and four side terminals (the rotating plug connects the central and two opposite side outlets); VD, electrical valve driver outside the compartment; TT, Teflon tubings and connectors; M, metallic box made from 3-mm brass plates; P, outer insulation of 20-mm polystyrene foam covered with aluminum foil; C and R, compensatory and respiratory vessels in one block of Plexiglas; L, closing lid with the outlets to respirometer; O, additional outlets used for air perfusion and for calibration.

The circuitry is painted as shown in Fig. 5.4 (P) and the rest of the metal is etched by $FeCl_3$ solution. The measuring membrane (M) is 40-μm beryllium-bronze foil (stainless steel can be also used), which is cemented by epoxy resin (Woodhill Chemical Sales Corp., Cleveland, Ohio) across the 12-mm hole in the central plate. The tensometric sensors (T) are placed on each side of the membrane. Their connecting wires are soldered to the printed circuits (P) and the whole plate, including the outer 1-mm rim of the membrane and the circuitry, is embedded into epoxy resin and cemented to a 2-mm Plexiglas plate (F, Fig. 5.4b). The whole transducer is assembled by adding two 3-mm brass plates (G),

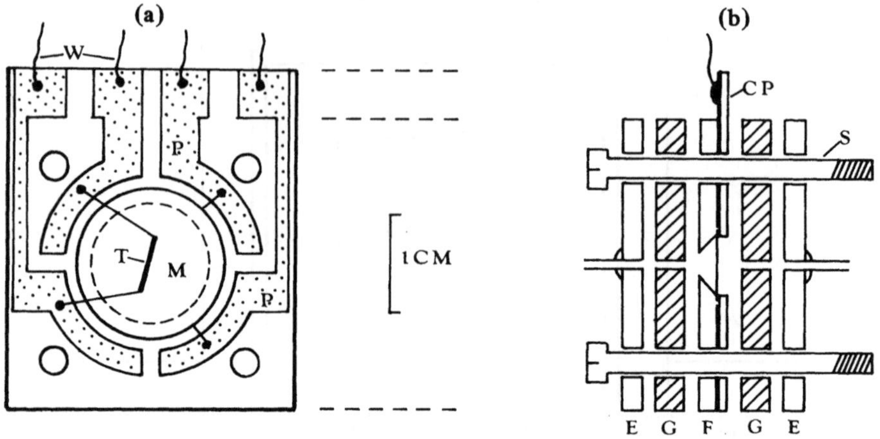

Fig. 5.4. Differential pressure transducer. **(a)** Front view of central plate of transducer. M, transducer membrane, 40-μm beryllium–bronze foil; T, semiconductor tensometer type AP-120-6-12; P, printed circuits with soldered joints of tensometers; W, connecting wires. **(b)** Lateral view of transducer. CP, central plate with membrane, sensors, and circuitry; F, 2-mm Plexiglas plate covering circuits and holding membrane; G, concentric 3-mm brass plates; E, outer 2-mm Plexiglas plates with stainless steel outlets, 0.5-mm o.d.; S, screws for assembling transducer.

one from each side of the central plate, and two 2-mm Plexiglas plates (E) with stainless steel outlets. The spaces between the plates are smeared with silicone grease and the plates are tightly pressed together by four screws (S) penetrating the concentric holes.

Various membranes (20–80 μm thickness) and mechanical strain gauge sensors have been tested for respirometric purposes. The small wire tensometers (type SSM-120, 1 × 1 mm grid size, 10-μm wire, 120 Ω; Mikrotechna Co., Prague, Czechoslovakia) showed outstanding temperature compensation effects, and perfect zero stability during long-term measurements, but they were not sensitive enough for work with very small tissue samples. Better sensitivity was achieved with semiconductor tensometers placed in the center of the transducer membrane on each side. They were connected to form one-half of the compensated resistance bridge—one recording positive and the other negative radial deformities of the membrane. The differential transducer shown in Fig. 5.4 had the membrane equipped with two tensometers of either type AP 120-6-12 or type AP 120-3-12 (OPS, Gottwaldov, Czechoslovakia). Their size was 6.0 × 0.6 × 0.02 mm or 3.0 × 0.2 × 0.01 mm, respectively, the deformation sensitivity factor K was +120 ± 5, and the resistance was 120 Ω. The equivalent semiconductor tensometers available in the United States are, for example, type SNB 2-06-12S9 or SN5-06-12S9 (BLH Electronics, Waltham, Mass.).

The tensometric sensors are fixed to the membrane by special glues of low hysteresis changes (cyanoacrylic glue or epoxy resins polymerizing at higher temperature). Before the transducer is assembled, the membrane, with the sensors, was properly dried and siliconized. The use of any hygroscopic glue for fixation of the sensors must be avoided, otherwise the transducer will serve also as a very sensitive hygrometer and slight changes in water vapor tension caused by KOH solution in the respirometer could cause instability in the zero position.

There is about 200 μl dead space between the transducer membrane and the inlet to the respiratory vessel. This can be decreased still further by filling the dead space with paraffin wax or by using connecting tubings of smaller diameter. However, it is not necessary as the instrument is able to record reliably less than 1 nl oxygen consumption changes with additional vessel volume of 200–1000 μl. The handling of the tissue samples, the incubation media, the CO_2 absorber, and other conditions for respirometry are the same as described for the volumetric respirometer in Sect. VII. The exception is that the respiratory vessels are not in the water bath. This enables easier manipulation of the respirometric samples.

Extensive recording with the instrument has demonstrated that the effects of temperature can be well compensated for in the gas phase, provided that the vessels are situated inside a heavy metal box that absorbs small changes in temperature regulation. The effects of temperature changes are also decreased by enclosing the whole volumetric unit in a 2-cm polystyrene cover packed in aluminum foil. The insulated instrument is then kept inside a larger compartment with an air thermostat regulating the temperature to within ±0.2°C.

B. Tensometric Unit

Electronic instruments for recording changes in mechanical quantities are commonly used in industry and research. They are high-precision oscillators and amplifiers designed to work with strain gauge, inductive, or capacitive sensors. An apparatus that can be used with the described semiconductor transducer, for example, is type CA-100 or type CA-110 (Peekel Instruments B.V., Rotterdam, Holland), or type KWS 73 (Hottinger-Baldwin Messtechnik GMBH, Darmstadt, Germany). The present work was done with the tensometric unit M-1000 (Mikrotechna Co., Prague, Czechoslovakia) consisting of five modules. One is common to all, supplying a constant stabilized DC current. It also releases the signals for automatic consecutive or simultaneous calibration of the electrical circuits and selects the voltage across the tensometric resistance bridges (1, 2, or 4 V). Each of the four remaining modules (M-1101) independently accommodates one transducer of the respirometer.

The module supplies a constant 5-kHz AC current to the tensometric resistance bridge. The resistance and capacitance across the bridge are balanced to zero, and any imbalance produced by mechanical deformation of the tensometers is amplified, decoded, and transformed into voltage or current changes of the DC output.

There are two independent output signals from each module. This makes it possible to record simultaneously the respirometric data from one module on two recorders set for different sensitivities and recording speeds. The sensitivity on the M-1101 module can be varied from 0.1 mV V^{-1} to 20 mV V^{-1}. In addition, each module has separate zeroing and amplifier gain controls so that the DC output signals of all four respirometers can be made equal to the same scale factor. The M-1101 module has one-half of the resistance bridge built-in (2 × 120 Ω) to facilitate the work with only two active tensometers.

C. Recorder

Any type of linear recorder with approximately 10-mV sensitivity can be used. To achieve automated respirometric scanning, a contact switch should be installed, indicating the upper limit of the recording scale. Some recorders are already equipped with this arrangement. The contacts of the switch are closed whenever the pen driver reaches the upper limit of the scale. This gives an electrical signal to the valve driver to start rotation of the valve. With the Hamilton 1XP valve driver the function of the contact can be described as follows: With 90° rotation the valve opens the respirometer for the first time, the pressure across the transducer membrane equilibrates to zero, the pen driver of the recorder drops to the initial position, and the contact is disconnected. The valve continues to rotate, opening the respirometer again at 270°C for fine zero adjustment before returning to the initial position indicated by the selector switch. The pen driver then repeats a new scanning until the contact is again connected.

This simple arrangement allows automatic performance of the respirometer, which is especially useful for prolonged measurements (hours or days). In this work I used the whole battery of VAREG recorders (Metra, Blanskò, Czechoslovakia) with sensitivity from 6 mV to 1.5 V or from 30 μA to 6 mA full scale. The recording speed could be varied from 20 to 3600 mm h^{-1}.

The recording on each respirometric module could be simultaneously monitored by an attached oscilloscope to reveal volumetric changes of higher frequency (ventilatory movements in insects). Finally, each of the recorders could be connected with the attached IT-2 electronic digital integrator (Laboratorni Pristroje, Prague, Czechoslovakia) for numerical data processing.

2. Calibration of the Respirograph

General information on the sensitivity of the transducer membrane can be obtained by applying known hydrostatic pressures from one side and recording the responses of the recorder. The transducer described above with a 40-μm beryllium-bronze membrane 12 mm in diameter, with two active semiconductor tensometers type AP 120-6-12 (1-V bridge voltage, 1 mV V^{-1} sensitivity range, 15 mV recorder scale) has shown a 400-mm pen driver deflection in response to a 1-Pa (about 0.1-mm H_2O) change of hydrostatic pressure.

With an increased sensitivity to the 0.1-mV V^{-1} range, the pen driver should exhibit a 4-mm deflection in response to relative hydrostatic pressure changes as small as 1 mPa. It is obvious that such enormous sensitivity would be useless to any absolute pressure transducer exposed to the atmosphere (frequent "noise" in atmospheric pressure is ±5 Pa or more). This sensitivity can be conveniently employed, however, when the transducer is used as a part of the closed differential respirometer.

Data on the sensitivity of the respirograph may be obtained by recording responses of the pen driver to known changes of volume in the respiratory vessel. This is achieved by means of a calibration syringe, which is either permanently connected to the outlets of the valve or occasionally attached to the respiratory vessel.

Theoretically, the relative pressure of air in a closed compartment at constant temperature is related to its volume according to the relationship $P_0 \cdot V_0 = P_1 \cdot V_1$, and it is a hyperbolic function. When the change of the volume $(V_1 - V_0)$ represents only a small fraction of the V_0 (1/1000 or less), the resulting change in relative pressure $(P_1 - P_0)$ will be almost equal to that produced by a consumption or formation of the same $(V_1 - V_0)$ amounts of air at a constant V_0 volume. This principle is used for calibration of the instrument. Thus, for instance, enlargement of the respiratory vessel volume by 1 μl (using the calibration syringe) should produce almost exactly the same response of the recorder as 1 μl of gas being consumed within the vessel.

In Fig. 5.5 the data are shown for respiratory vessels with a 1-10 ml capacity, measured at a medium sensitivity of 1 mV V^{-1}. The hyperbolic relationship is shown: for example, a 1-μl change in the volume of the respiratory vessel of 1-ml capacity causes about a 200 mm pen driver reaction. For the sensitivity range of 0.1 mV V^{-1} this would indicate a 2 mm response to a 1-nl volume change in a 1-ml vessel. For work with small tissue samples this sensitivity is further increased to more than a 10 mm response for a 1-nl change in gas volume simply by using vessels of smaller than 1-ml capacity. Moreover, the sensitivity can be further increased 10 times using a 4 V instead of a 1 V bridge voltage and a 6 mV instead of 15 mV recorder scale. This is often associated with zero instability and increased electrical noise level. With the described transducer the recorder response was linear for all physiological ranges of oxygen consumption to be measured. The membrane has been tested to hydrostatic pressures of up to 5 kPa with no influence on the electrical properties or zero position.

The last and most practical method of calibration may be performed during any respirometric experiment. The calibration syringe is used as a compensatory device in the same way as described for the volumetric respirometer in Sect. VII. To this end, changes on the recorder (produced by the oxygen consumption of the tissue sample present in the respiratory vessel) are compensated to zero by means of the calibration syringe (Hamilton 7001). The scale factor on the recorder is then directly determined from the compensating volume. The possibility of such a direct correlation of electrical recordings with volumetric data increases the versatility of this respirometer.

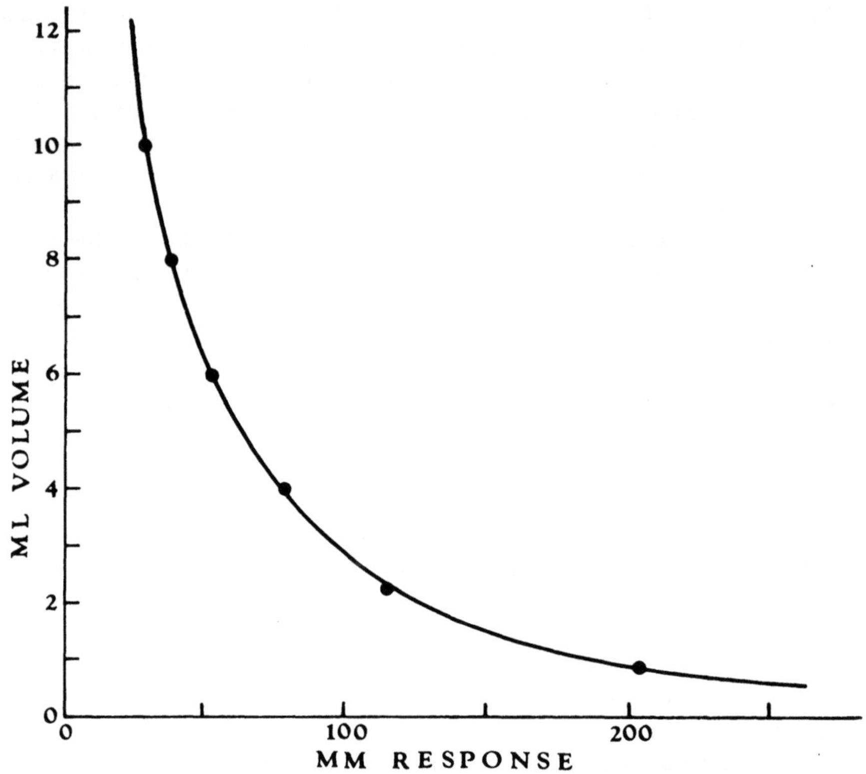

Fig. 5.5. Dependence of sensitivity of differential pressure transducer on volume of respiratory vessels (compensatory and respiratory vessels of equal capacity). Values indicate responses of pen driver to 1-μl increase of volume in respiratory compartment at 25°C, 1-mV V^{-1} sensitivity range. The change recorded with 1-ml vessel corresponds approximately to a 500-mPa decrease in relative pressure.

Occasionally it is necessary to check for possible leakage of the joints. This is made by applying positive or negative hydrostatic pressures (20–50 mm H_2O) to the central outlet of the opened respirometer valve. When the valve is closed the recorder immediately shows rapid changes of pressure due to equilibration with the atmospheric pressure on the leaking side. Another way to identify leakage in one of the respirometric compartments is to close the valve and increase the sensitivity. The leakage will be immediately identified by the appearance of an atmospheric "noise" of pressure and, in addition, the instrument will record movements of the door or window.

3. Temperature Compensation

The differential manometric principle of the respirograph ensures that small temperature changes acting simultaneously on both the respiratory and compensatory compartments are eliminated. It appears that temperature changes can be sufficiently compensated for when the respiratory vessels are closed in a solid metallic block that buffers any sudden changes in temperature. With vessels smaller than 1 ml made from plastic materials, the 15-min period was sufficient for temperature equilibration.

Electrical compensation of temperature changes is achieved by selection of identical pairs of semiconductor tensometers with equal temperature coefficients. They are connected in neighboring branches of the resistance bridge so that temperature changes of equal polarity should be theoretically equilibrated to zero. Because of some slight changes in the electrical properties of the sensors after their fixation to the membrane, a 100% temperature insensitivity is practically never obtained. Since all the measurements are made at a constant temperature, there are no problems with the temperature influence at medium or lower sensitivity ranges. With a maximum sensitivity of 0.1 mV V^{-1} there is occasionally slight zero drift, apparently due to small changes in temperature. However, this zero drift can be easily measured and accounted for by recording with the opened respirometer valve. Because the transducer and the valve are located in a metallic box insulated from the outside by polystyrene and aluminum foil, it takes about 6 h for perfect temperature equilibration in the transducer. For this reason the whole volumetric unit of the respirograph is permanently kept at 27°C.

4. Working Procedure

A. Dissect the tissue and measure its volume as described in Sects. V and VI.
B. Transfer the selected tissue (30–250 nl) into the incubation medium (1–2 μl) and place it along with the strip of porous plastic containing osmotically balanced KOH solution into respiratory vessels of 200–1000 μl capacity. Place medium without tissue plus KOH into a compensatory vessel of equal or slightly larger capacity (see Sect. VII).
C. Open the thermostatic box and the insulated box housing the respiratory vessels and connect the vessels to the outlets of the transducer; the respirometer valve is opened.
D. Balance the bridge current to the zero position of the indicator on the tensometric unit; use 1 V bridge voltage. Select the required sensitivity range and adjust the zero position of the recorder.
E. Allow 15–20 min for temperature equilibration; during this time the pen driver records electrical zero stability at the given sensitivity range.
F. If new vessels are used, perform calibration of the instrument according to instructions given in Sect. VIII.2.
G. Turn the selector switch of the valve to the "closed" position and connect the circuit for automatic valve rotation when the limit of the scale has been attained. The respirometer will then perform automatic scanning of oxygen consumption at selected sensitivity and scan speed (Fig. 5.6). When larger sensitivity is required, turn the selector switch to the "open" valve position, change the sensitivity, adjust the zero position of the recorder, and close the valve again.
H. For exchange of the tissue sample, open the valve, exchange respiratory vessels, and repeat steps F and G.

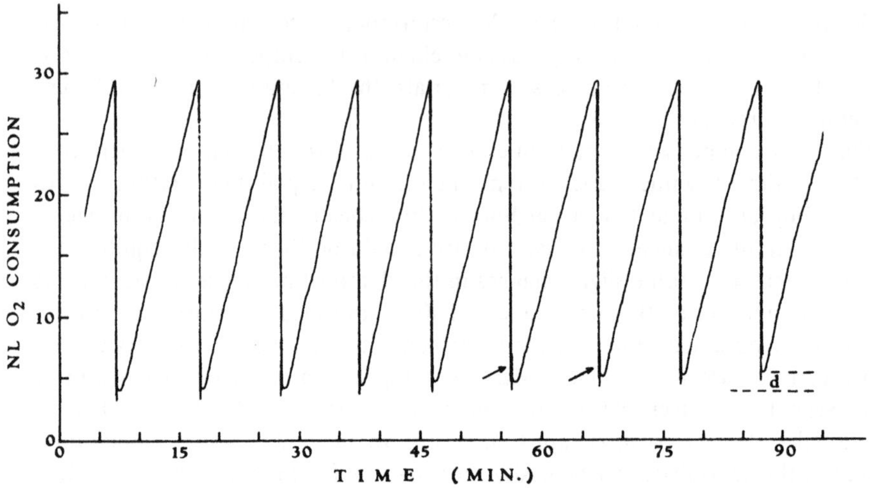

Fig. 5.6. Trace record from scanning microrespirograph showing oxygen consumption of 251 μg intestinal epithelium of adult female *Pyrrhocoris apterus*. Epithelium was incubated in 1.5 μl Grace's tissue culture medium in 200-μl respiratory vessels at 27°C. Respiration rate is oscillating around 850 nl O_2 h^{-1} mg^{-1}. Arrows, moments of automatic valve openings; d, electrical zero drift for entire period.

The tensometric units M-1000 (Mikrotechna) or CA-110 (Peekel Instruments, B.V.) makes it possible to use six or more separate respirographs. In work with tissue culture it is more practical, however, to use only one respirograph with a common compensatory vessel and up to 10 respiratory vessels situated in one holder. After temperature equilibration each of the respiratory test tubes is consecutively connected to the transducer by means of Hamilton miniature Teflon valves and joints. Oxygen consumption is sequentially scanned for 10–20 min in each tube. When incubated in the proper tissue culture media, various insect epithelia exhibit rather constant rates of oxygen consumption over prolonged periods of time. An example obtained with the intestinal epithelium is shown in Fig. 5.6.

References

Bernardini PM, Laudani U (1966) La consommation d'oxygène de la glande prothoracique de *Leucophaea maderae* et *Periplaneta americana* d'après l'aspect histologiques des organes endocrines. J Insect Physiol **12**:1289–1294

Bodine JH (1950) To what extent is oxygen uptake of the intact embryo related to that of its homogenate? Science **112**:110–111

Bodine JH, Lu KH (1950) Oxygen uptake of intact embryos, their homogenates and intracellular constituents. Physiol Zool **23**:301–308

Bodine JH, West WL (1953) Respiratory quotients of intact egg, isolated embryo and embryo homogenate. Iowa Acad Sci **60**:594–598

Brown JJ, Chippendale GM (1977) Ultrastructure and respiration of the fat body of diapausing and non-diapausing larvae of the corn borer, *Diatraea grandiosella*. J Insect Physiol **23**:1135–1142

Buck J (1962) Some physical aspects of insect respiration. Annu Rev Entomol **7**:27–56

Crisp DJ, Thorpe WH (1947) A metal micro-respirometer of the Barcroft type suitable for small insects and other animals. J Exp Biol **24**:304–309

Degn H, Lundsgaard JS, Petersen LC, Ormicki A (1980) Polarographic measurement of steady state kinetics of oxygen uptake by biochemical samples. Methods Biochem Anal **26**:47–77

Dixon M (1951) Manometric methods as applied to the measurement of cell respiration and other processes, 3rd ed, Cambridge University Press, New York

Fourche J (1964) Un respiromètre électrolytique pour l'étude des pupes isolées de Drosophile. Bull Biol Fr Belg **48**:475–489

Fourche J, Ambrosioni JC (1969) Le métabolisme respiratoire au cours des métamorphoses, respiration in vitro des ovaires de *Bombyx mori*. J Insect Physiol **15**:1991–1997

Gilby AR, Rumbo ER (1980) Water loss and respiration of *Lucilia cuprina* during development within the puparium. J Insect Physiol **26**:153–162

Gilson WE (1963) Differential respirometer of simplified and improved design. Science **141**:531–532

Glick D (1949) Techniques of Histo- and Cytochemistry. Interscience, New York

Hamilton AG (1959) The infra-red gas analyser as a means of measuring the carbon dioxide output of individual insects. Nature **184**:367–369

Hamilton AG (1964) The occurrence of periodic or continuous discharge of carbon dioxide by male desert locusts (*Schistocerca gregaria* Forskal) measured by infra-red gas analyser. Proc R Soc Lond (Biol) **160**:373–395

Jones JC (1977) The circulatory system of insects. Thomas, Springfield, Ill.

Keeley LL, Friedman S (1967) Corpus cardiacum as a metabolic regulator in *Blaberus discoidalis* Serville (Blattidae) I. Long-term effects of cardiacectomy on whole body and tissue respiration and on trophic metabolism. Gen Comp Endocrinol **8**:129–134

Keister M, Buck J (1974) Respiration: Some exogenous and endogenous effects on rate of respiration. In: Rockstein M (ed) The physiology of Insecta, 2nd ed, pp 469–509. Academic Press, New York

Kleinzeller A (1965) Manometrische Methoden und Ihre Anwendung in Biologie und Biochemie. Fischer, Jena, GDR

Klekowski RZ, Zajdel JW (1972) Capacity electrolytic respirometer KZ-CER-OIT with review and discussion of electrolytic respirometry. Pol Arch Hydrobiol **19**:475–504

Kuusik A (1976) Cyclic gas exchange in adult Coleoptera studied by continuous gas-chromatographic registration. Izv Akad Sci Estonian SSR **25**:97–105

Kuusik A (1977) Cyclic gas exchange in diapausing pupae of *Pieris brassicae* L. and *P. rapae* L. (Lepidoptera, Pieridae). Izv Akad Sci Estonian SSR **26**: 96–101

Leenders HJ, Knoopien WG (1973) Respiration of larval salivary glands of *Drosophila* in relation to the activity of specific genome loci. J Insect Physiol **19**:1793–1800

Lessler MA, Brierley GP (1969) Oxygen electrode measurements in biochemical analysis. Methods Biochem Anal **17**:2–29

Ludwig D, Barsa MC (1956) Oxygen consumption of whole insects and insect homogenates. Biol Bull **110**:77–82

Ludwig D, Barsa MC (1957) Respiratory metabolism of homogenates during the embryonic development of the mealworm, *Tenebrio molitor* Linnaeus, with added substrates and inhibitors. Ann Entomol Soc Am **50**:475–477

Lüscher M (1968) Hormonal control of respiration and protein synthesis in the fat body of the cockroach *Nauphoeta cinerea* during oocyte growth. J Insect Physiol **14**:499–511

Mill PJ (1974) Respiration: Aquatic insects. In: Rockstein M (ed) The physiology of Insecta, 2nd ed, pp 403–467. Academic Press, New York

Miller PL (1974) Respiration-aerial gas transport. In: Rockstein M (ed) The physiology of Insecta, 2nd ed, pp 345–402. Academic Press, New York

Müller HP, Engelmann F (1968) Studies on the endocrine control of metabolism in *Leucophaea maderae* (Blattaria) II. Effect of the corpora cardiaca on fatbody respiration. Gen Comp Endocrinol **11**:43–50

Oberlander H (1980) Tissue culture methods. In: Miller TA (ed) Cuticle techniques in arthropods, pp 235–272. Springer, New York

Punt A (1950) The respiration of insects. Physiol Comp **2**:59–74

Punt A (1956) Further investigations on the respiration of insects. Physiol Comp **4**:121–131

Putman RS (1976) The gas chromatograph as a respirometer. J Appl Ecol **13**: 445–452

Sacktor B (1974) Biological oxidations and energetics in insect mitochondria. In: Rockstein M (ed) The physiology of Insecta, 2nd ed, pp 271–355. Academic Press, New York

Samuels A (1956) The effect of sex and allatectomy on the oxygen consumption of the thoracic musculature of the insect, *Leucophaea maderae*. Biol Bull **110**:179–183

Scholander PF (1942) Volumetric microrespirometers. Rev Sci Instrum **13**: 32–33

Scholander PF (1950) Volumetric plastic microrespirometer. Rev Sci Instrum **21**:378–380

Scholander PF, Iversen O (1958) New design of volumetric respirometer. Scand J Clin Lab Invest **10**:429–431

Scholander PF, Claff CL, Andrews JR, Wallach DF (1951) Microvolumetric respirometry. J Gen Physiol **35**:375–395

Sláma K (1965) Effect of hormones on the respiration of body fragments of adult *Pyrrhocoris apterus* L. (Hemiptera). Nature **205**:416–417

Tadmor U, Applebaum SW, Kafir R (1971) A gas chromatographic micromethod for respiration studies on insects. J Exp Biol **54**:437–441

Taylor P (1977) A continuously recording respirometer, used to measure oxygen consumption and estimate locomotor activity in tsetse flies, *Glosina morsitans*. Physiol Entomol **2**:241–245

Tobias JM (1942) Membrane interferometer manometer. Rev Sci Instrum **13**: 232–233

Tobias JM (1943) Microrespiration techniques. Physiol Rev **23**:51–75

Umbreit WW, Burris RH, Stauffer JF (1972) Manometric and biochemical techniques, 5th ed. Burgess, Minneapolis

Wang CH (1967) Radiorespirometry. Methods Biochem Anal **15**:312–368

Wiens AW, Gilbert LI (1965) Regulation of cockroach fat-body metabolism by the corpus cardiacum in vitro. Science **150**:616–617

Wightman JA (1977) Respirometry techniques for terrestrial invertebrates and their application to energetics studies. NZ J Zool **4**:453–469

Winteringham FPW (1959) An electrolytic respirometer for insects. Lab Pract **8**: 372–376

Chapter 6
Spectrophotometry and Fluorometry in Ion Transport Epithelia

Lazaro J. Mandel

I. Introduction

The subject of this chapter is the use of spectrophotometry and fluorometry to monitor the redox state of respiratory enzymes in intact epithelial tissues, with an emphasis on insect epithelia. These optical methods were used initially in isolated mitochondria and later in a variety of intact tissues *in vitro* and even *in vivo*. Their main advantages reside in being rapid and noninvasive, the latter allowing the optical monitoring to be performed in conjunction with other types of measurements normally used to assess epithelial function. Spectrophotometry and fluorometry of intact epithelia have provided novel information not obtainable by other methods.

In the next section some basic principles concerning the function of the mitochondrial respiratory chain are discussed and the concept of mitochondrial transitions is introduced. A detailed description is then given of the optical methods used, especially the design of different types of epithelial chambers. The remainder of the chapter is devoted to the various uses of fluorometry and spectrophotometry in epithelia.

II. Mitochondrial Respiratory Chain

The respiratory chain consists of a series of hemoproteins that transfer electrons from various reactions of intermediate metabolism to oxygen. The entire chain is confined to the mitochondria, where the components are bound to the inner membrane. Most of the information regarding the functioning of the respiratory chain has been obtained from studies with isolated mitochondria. A schematic

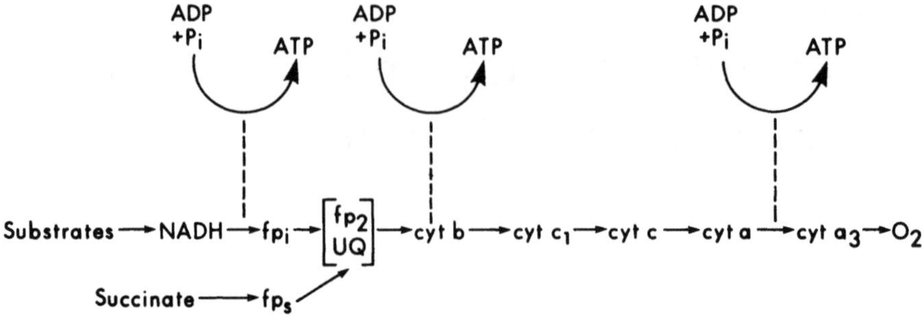

Fig. 6.1. Mitochondrial respiratory chain. cyt, cytochrome; fp, flavoprotein(s); UQ, ubiguinone.

representation of the flow of electrons through the respiratory chain and their coupling to oxidative phosphorylation is shown in Fig. 6.1. Reducing equivalents from various metabolic substrates enter the respiratory chain through pyridine nucleotide (NADH)- or flavoprotein (fp)-linked reactions and continue as electron movement through the chain to be oxidized by oxygen. In the process, electronic potential energy is converted into biochemical energy in the form of adenosine triphosphate (ATP). Chance and Williams (1956) described five respiratory states for isolated mitochondria and the conditions required to elicit a change from one state to another. These states represent extremes of mitochondrial function not necessarily achievable in intact tissues; nevertheless, these states provide an excellent reference point from which to understand mitochondrial properties. The mitochondrial respiratory states are defined and described in Table 6.1.

State 1 characterizes the mitochondria when they are first isolated, containing low levels of substrate and adenosine diphosphate (ADP) and exhibiting a low respiratory rate. State 2 is called "starved" because the mitochondria are provided with sufficient oxygen and ADP but are deprived of metabolic substrates. In the extreme case of total substrate deprivation shown here, there is no flow of reducing equivalents into the mitochondrial chain and, thus, all the respiratory components are oxidized. State 3 is the "active" one because it describes the conditions present when the mitochondria are respiring at their maximal rate in the presence of sufficient substrate, ADP (and inorganic phosphate, P_i), and oxygen. Note that the respiratory chain components display a progressively increasing level of reduction, starting with cytochrome a, a_3, which is almost completely oxidized, to nicotinamide adenine dinucleotide (NAD), which is about 50% reduced. State 4 is the "resting" state, achieved by depletion of ADP or P_i, and characterized by a low respiratory rate and generally a more reduced condition than state 3. State 5 is achieved upon anoxia and causes all the components to become totally reduced.

It is simple to visualize mitochondrial behavior during transitions between states. For example, the transition between state 4 (resting) and state 3 (active)

Table 6.1. Respiratory States of Mitochondria[a]

State	ADP, P_i	Substrate	O_2	qO_2	Percentage reduced				Rate limitation
					NAD	Cyt *b*	Cyt *c*	Cyt *a,a*$_3$	
1	Low	Low	High	Slow	90	17	7	0	ADP
2 (starved)	High	0	High	Slow	0	0	0	0	Substrate
3 (active)	High	High	High	Fast	53	16	6	4	Respiratory chain
4 (resting)	Low	High	High	Slow	99	35	14	0	ADP
5 (anoxic)	High	High	0	0	100	100	100	100	Oxygen

[a] Adapted from Chance B, Williams CM (1956) The respiratory chain and oxidative phosphorylation. Adv Enzymol **17**:65–134.

is usually achieved by the addition of ADP. The respiratory rate accelerates rapidly and the respiratory enzymes become more oxidized, with the exception of cytochrome a, a_3. The reverse would happen in a transition from the active to the resting state. In intact tissues, the extremes represented by states 3 and 4 are not usually observed (see Sect. IV); rather, transitions to a "more active state" or "more resting state" are obtained by increasing or decreasing cellular work, respectively. On the other hand, the transition from either state 3 or 4 (or any state between them) to state 5 (anoxia) elicits the complete reduction of all respiratory enzymes in isolated mitochondria and intact tissues.

In the transition between states 3 and 4, the enzyme component that exhibits the most dramatic shifts in redox level is NAD, being 99% reduced in state 4 and 50% oxidized in state 3. Mitochondrial NAD is associated with the initial step of the mitochondrial respiratory chain. Thus, the steady-state redox level of NAD is a function of both its rate of oxidation by the flavoproteins and its rate of reduction by the numerous dehydrogenases of intermediary metabolism. By monitoring the redox state of mitochondrial NAD (see below), the relative rates of these two categories of redox reactions can be analyzed, thereby providing important information on the respiratory state of the mitochondria.

Conversely, in the transition to anoxia, cytochrome a, a_3 displays the largest change in redox level. In fact, this cytochrome has such a high affinity for oxygen that it remains fully oxidized as long as the local oxygen concentration is above 1 μM (Oshino et al. 1974). In isolated mitochondria the transition between full reduction of cytochrome a, a_3 and almost full oxidation occurs between 10^{-8} and 10^{-6} M, with a half-point at about 10^{-7} M, which is less than 0.1 mm Hg. At higher oxygen levels, these curves become independent of partial oxygen pressure (PO_2). Such a high affinity for oxygen makes the redox level of this cytochrome an excellent probe for tissue anoxia (see below); reduction in this cytochrome indicates that the oxygen level is below 10^{-7} M. From Table 6.1 it may also be seen that the redox levels of the other cytochromes change with respiratory state; thus, cytochrome redox state may also be used to monitor mitochondrial state transitions.

III. Methods

1. Fluorometry

Fluorometry has been used extensively to monitor the redox level of mitochondrial NADH. Reduced NAD (NADH) fluoresces in a broad emission band of 425 to 500 nm when excited with 310-370 nm light, whereas the oxidized form of NAD (NAD^+) does not fluoresce. Accordingly, the intensity of fluorescence at 450 nm is proportional to the concentration of NADH. Binding of NADH to the inner mitochondrial membranes causes a fluorescence enhancement of an order of magnitude (Avi-Dor et al. 1962). This property has permitted the measurement of the redox state of NAD with great sensitivity in isolated mitochondria and intact tissues by fluorometric and microfluorometric techniques.

The use of fluorometry in intact tissues was pioneered by Chance and Jöbsis (1959) and subsequently modified by them and their co-workers (Chance et al. 1962; Jöbsis and Duffield 1967). The most commonly used methods have been described by Jöbsis and Stainsby (1968) and Rosenthal and Jöbsis (1971). An arrangement used by Balaban et al. (1981) for microfluorometry of NADH in isolated perfused proximal tubules is shown in Fig. 6.2, which serves as an excellent illustration for the use of fluorometry in general.

Light from a mercury arc lamp is sharply filtered to obtain a narrow band of excitation light centered around 366 nm. The incident light may be applied to the tissue either directly or through light fibers. The angle between the incident

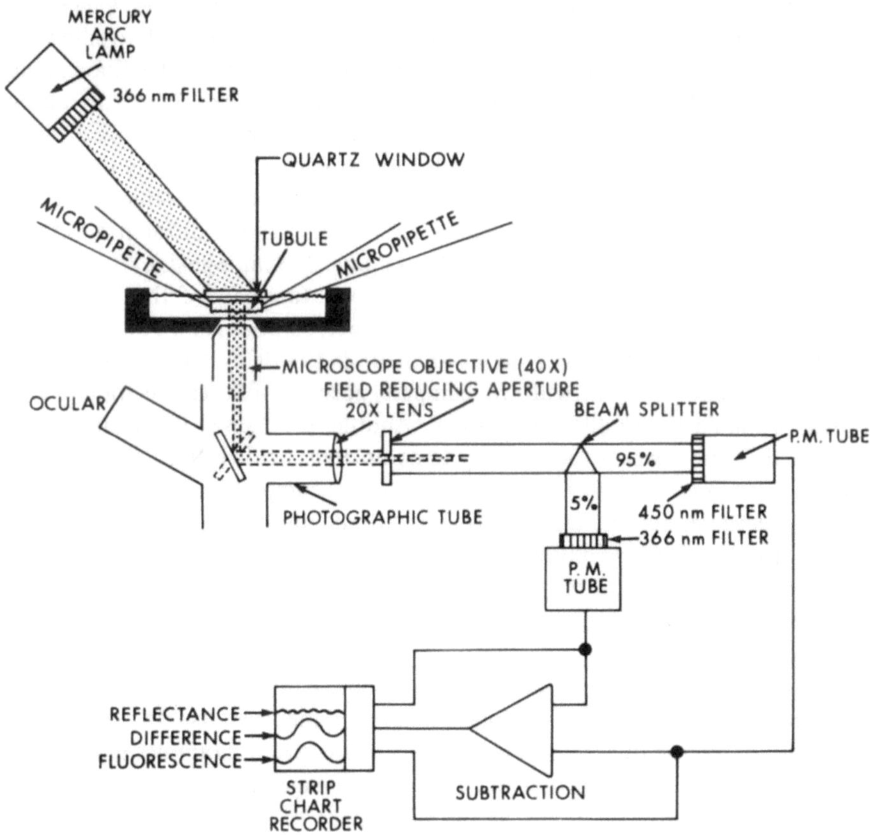

Fig. 6.2. Apparatus used for microfluorometry of isolated perfused tubules. [From Balaban RS, Dennis WW, Mandel LJ (1981) Microfluorometric monitoring of NAD redox state in isolated perfused renal tubules. Am J Physiol **240**: F337–F342]

light and the detector must be selected so as to minimize specular reflection from the tissue. This item is crucial for measurements of surface fluorescence (not shown) or transmission fluorescence (as shown in Fig. 6.2). Conversely, for fluorescence measurements in a tissue suspension (see Figs. 6.8 and 6.9), the amount of light scattering is so large that the angle between the light input from the mercury arc lamp and the light output to the photomultiplier tubes is not crucial. Good results were obtained in the chamber shown in Fig. 6.7, which has a straight line between input and output.

For the microfluorometric configuration shown in Fig. 6.2, a standard tubule perfusion apparatus is used, as modified by Balaban et al. (1981) for fluorescence measurements. The combined emitted and reflected excitation light from the tissue is directed to the photomultiplier tubes through the photographic output tube of the microscope. In general, when fluorometric measurements are performed on a larger tissue sample, the output light may be collected with just a regular microscope objective in front of the photomultiplier tubes. One photomultiplier tube receives 95% of the output light and is filtered to allow the passage of only 450–490 nm light; this is considered to be the fluorescence signal from NADH. The other photomultiplier tube receives 5% of the output light and is filtered to allow only 366-nm light to pass; this is termed the reflected light and is used to compensate for variability in the intensity of the light source, scattering changes in the tissue, and to indicate movement artifacts (Jöbsis and Stainsby, 1968). All signals are reported as the difference between tissue fluorescence and reflectance (450-nm minus 366-nm signals).

The absolute magnitude of the signals are a function of the background fluorescence (i.e., the fluorescence of the chamber, solution, and 366-nm filters), the excitation angle, and the amount of tissue in the field. Because of this complex interaction, it is difficult to compare absolute fluorescent values from tubule to tubule and to different systems. Another problem in the quantification of NADH fluorescence is the inability to completely oxidize NADH in the intact tissue (Balaban et al. 1982). As discussed earlier, complete reduction of NAD is easily achievable, but the lack of a fully oxidized condition impedes the delineation of a reduced minus oxidized scale for NADH. Such a scale would allow the direct calculation of the mitochondrial NADH/NAD ratio, a variable that appears to control numerous tissue metabolic functions (Balaban et al. 1982; Hansford 1980).

Thus, changes in NADH fluorescence can only be interpreted qualitatively or semiquantitatively. With these limitations in mind, we use in general two methods to express fluorescence values: (1) In qualitative examples in which the direction and not the magnitude of fluorescent changes is important, the fluorescent scale is reported in terms of the percentage of change in absolute tissue fluorescence. (2) When quantification is required, a fluorescent scale is defined for each tubule as the range of values obtained between control conditions (0%) and those observed in the presence of 0.1 m*M* KCN or anoxia (100%). The change in fluorescent emission that occurs with each experimental condition is quantified with the use of this scale, each tubule being used as its own internal control.

2. Spectrophotometry

The use of spectrophotometry to monitor the redox state of respiratory enzymes is based on their well-known property of absorbing light at characteristic wavelengths when in the reduced form as compared to their oxidized form (except for flavoproteins, which absorb in the oxidized form). A typical reduced-minus-oxidized spectrum of a rat liver mitochondrial suspension is shown in Fig. 6.3. The spectrum is divided into α, β, and γ bands since the cytochromes absorb light in three different regions of the spectrum. Starting from the right, it shows the cytochrome a, a_3 α peak at 605 nm, the cytochrome b α peak (shown as a shoulder) at 564 nm, the cytochrome $c(+c_1)$ α peak at 550 nm, constituting the α absorption bands. These cytochromes also absorb in the β region, but their peaks are indistinguishable. Continuing to the left, the absorption scale is 2.5 times larger because light absorption is more intense in this region. First, there is a trough at 465 nm due to the absorption of oxidized flavoproteins in the reference cuvettes. The cytochrome a_3, a γ peak at 445 nm and cytochrome b γ at 430 nm complete the identifiable absorption spectrum of the mitochondrial cytochromes. After another trough at 410 nm due to oxidized peaks in the reference cuvette, the NADH absorption peak at 340 nm is identified.

The reduced-minus-oxidized spectrum shown in Fig. 6.3 represents the maximum absorption of the respiratory chain components, and it is used as a

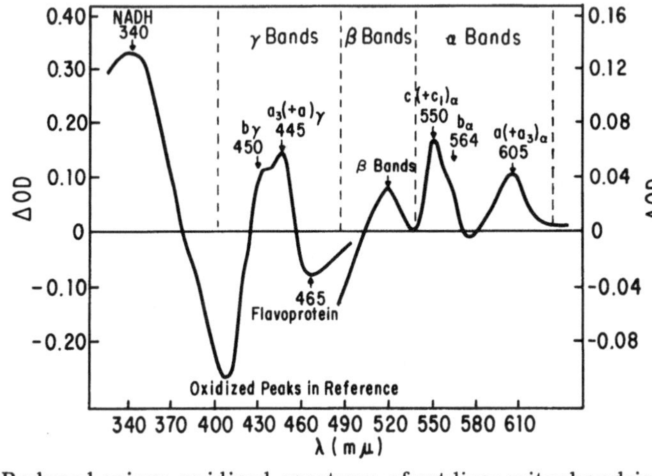

Fig. 6.3. Reduced minus oxidized spectrum of rat liver mitochondria. [Adapted from Jöbsis FF (1963) Spectrophotometric studies on intact muscle. I, II. J Gen Physiol **46**:905–969]

reference against which to measure the smaller changes that occur during other functional transitions. It is well known that absorption within each band is directly proportional to the concentration of the reduced form of each cytochrome, allowing the quantification of cytochrome redox state (Chance and Williams 1956). Similar absorption spectra have been obtained in a number of intact tissues, including epithelia (see below); thus, the same types of analyses may be used in these tissues.

The techniques presently used for the spectrophotometry of mitochondria and intact tissues were developed largely by Chance and co-workers (for review see Chance 1957). The measurement of changes in the absorption bands of the cytochromes–in the presence of tissue samples that exhibit changes of their light-scattering properties during the course of an experiment–required the development of techniques that compensate for such nonspecific changes in light absorption. Two such basic techniques were developed by Chance and his colleagues: (1) differential wavelength scanning (the "split beam"), in which the light from a monochromator falls alternatively on two samples and the differences in optical absorption are recorded as a "difference spectrum" (Yang 1954; Yang and Legallais 1954); and (2) dual-wavelength differential ("the double beam"), in which two beams of light at different wavelengths alternately fall on a single sample and the differences in optical absorption measure the redox state of a respiratory enzyme as a function of time (Chance 1951, 1954).

In the first method, the two tissue samples are identical, and thus one sample can be used as reference to subtract nonspecific absorption changes. The region of the spectrum of interest is scanned either manually or via motor control to record the difference in optical density between the two samples. An example is shown in Fig. 6.3, in which the reference sample is fully oxidized in the presence of oxygen and no substrates (state 2), and the other sample is fully reduced in the absence of oxygen but presence of substrates (state 5). In Fig. 6.3, the greater absorption in the anoxic state over the oxidized state is plotted in the upward direction. This method is excellent for recording differences in absorption occurring during steady states. For rapid changes in absorption, it is limited by the scanning speed, which can be as slow as 5 min for the entire visible spectrum, depending on the spectrophotometer.

In the second method, the wavelength of one beam is selected to correspond with the absorption peak of the component of interest; for the other beam a nearby wavelength is selected to act as reference. By monitoring of the difference in optical density between these two wavelengths, nonspecific absorption changes are minimized, since the latter are only slowly affected by wavelength. The wavelength pairs used for double-beam spectrophotometry of mitochondrial cytochromes have been selected from spectra similar to the one shown in Fig. 6.3. The values most commonly used are shown in Table 6.2. Since the two wavelengths are continuously being monitored, this method allows the measurement of rapid kinetic changes in redox state. Its main advantage is that only one respiratory chain component can be observed at a time. Spectrophotometers

Table 6.2. Peak and Reference Wavelengths Most Commonly Used in Double-Beam Spectrophotometry

Component	Peak wavelength (nm)	Reference wavelength (nm)
Cytochrome a_3,a	445	455
Cytochrome a,a_3	605	630
Cytochrome b	564	575
Cytochrome c	550	540
Flavoprotein	465	510
NADH	340	374

Data from Chance and Williams (1956).

of more recent design have partially overcome this difficulty by using four or more beams of different wavelengths to monitor two or more components simultaneously.

A third approach, the rapid-scanning spectrophotometer, has become feasible only recently due to advances in detector technology. This instrument produces results that may be deemed a combination of the two aforementioned methods (Mandel et al. 1976) since it is capable of monitoring a wide spectrum (340–640 nm, for example) with good resolution and rapid data collection. A block diagram of the instrument is shown in Fig. 6.4. Light from a 150 W xenon lamp is used to illuminate the sample either directly or through light fibers. The sample is held in one of the chambers described below (see Figs. 6.5–6.7), which permit simultaneous transillumination and measurement of other tissue properties. A light fiber bundle is used to transmit the light exiting from the tissue to the entrance silt of a monochromator. This monochromator is designed with two gratings mounted back-to-back on a swivel post for ease in selecting the appro-

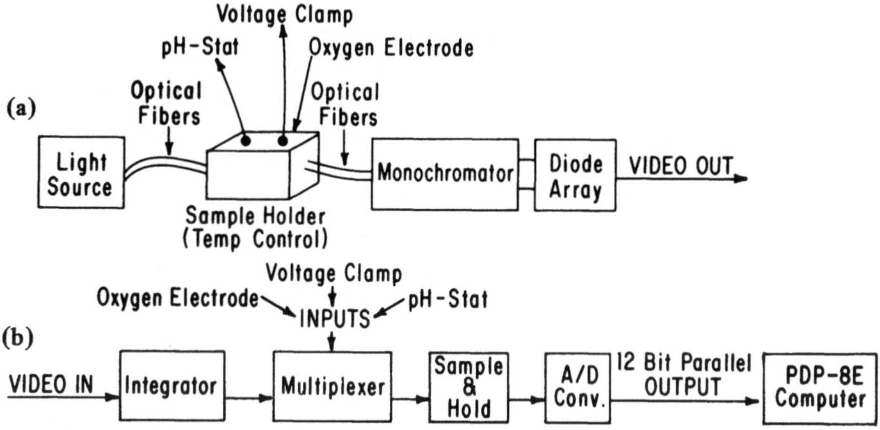

Fig. 6.4. Rapid-scanning spectrophotometer.

priate grating. These gratings have 295 and 147.5 lines/mm, respectively, allowing them to diffract the incident light in a narrow angle and in this manner project an image of the spectrum of either 150 or 300 nm within a width of approximately 1 cm at the monochromator exit optics.

In an earlier design, a television camera with an image-intensified vidicon was mounted flush against the monochromator output to scan the spectrum with 512 horizontal video lines. A more recent design makes use of a spectroscopic diode array, cooled to −20°C, as the light detector. The detector scans the diffracted output of the monochromator, dividing it into 512 equal lines, each producing an electrical signal that is proportional to the light intensity of a very narrow portion of the spectrum. Theoretical resolutions of 0.75 and 0.37 nm/line are obtained, depending on the grating used. In practice, however, the resolution is limited by the light-scattering properties of the tissue and the optical aberration of the monochromator.

The various manipulations that the electrical video signal undergoes to be interfaced with a digital computer are shown in Fig. 6.4b. (For a more detailed description of this type of interfacing see Joyner and Moore 1973.) The electrical output from each line is integrated as the camera scans that line. The output from the integrator passes through a multiplexer to a sample-and-hold circuit, and thereafter is converted into digital form in an analog-to-digital converter. The digital signal from each line is stored in a specified buffer location in a PDP-8E computer. The same procedure is repeated for each of the 512 lines; therefore, at the end of one complete scan of the whole spectrum, which lasts 33 msec for the television camera and is variable for the diode array, the amplitudes of 512 separate points in the output spectrum are stored in the computer buffer. Multiple scans can, of course, be averaged to improve the signal-to-noise ratio.

The multiplexer allows signals from other instruments to be read into the computer at the beginning of each scan; 40 locations are available for these signals. A scan, or average scan, under a specified set of conditions is stored on magnetic tape. In this manner, consecutive spectra are measured and stored as the experimental conditions are changed to obtain kinetic information about the properties of the sample under investigation. At the end of the experiment, whole spectra may be subtracted from each other to obtain difference spectra under two different sets of experimental conditions, or the redox state of each of the components of the cytochrome chain (with optical absorption within the measured range) may be simultaneously followed kinetically as the experimental conditions are changed. The spectra or the kinetic results are displayed on a graphics terminal that is attached to a copier to provide permanent records. An analog output has also been designed into the system to obtain on-line or off-line traces on a multiple-channel recorder.

It should be noted that fundamental differences exist between the designs and performance of the conventional as compared to the rapid-scanning spectrophotometers: (1) In split-beam or double-(multiple)-beam spectrophotometers, wavelength selection is performed prior to the light entering the sample and thus monochromatic light passes through the sample. A separate monochromator is required for each wavelength used. In rapid-scanning spectrophotometry, the wavelength selection is performed after the light passes through the tissue and thus white light is used for transillumination of the sample. Only one monochromator is required in this instrument. (2) The sensitivity of the detectors used is very different. Split-beam and double-(multiple)-beam spectrophotometers use photomultiplier tubes, or even photon-counting systems, as detectors. This feature makes them amenable for use under extremely low light levels. On the other hand, rapid-scanning spectrophotometers use solid-state diodes, which at the present state of the art cannot match the sensitivity of photomultipliers.

3. Chamber Design

The noninvasive nature of the optical measurements described in this chapter make them particularly amenable to the simultaneous use of other techniques to measure various aspects of tissue function. The parameters to be measured (i.e., short-circuit current, oxygen consumption, etc.), the constraints of the tissue to be studied, and the experimental apparatus all combine to dictate chamber design. Some examples of chamber types used for combined optical and other functional measurements are given below.

The chamber shown in Fig. 6.5 was designed to perform optical studies on sheet epithelia (Mandel et al. 1975) using a split-beam spectrophotometer, which

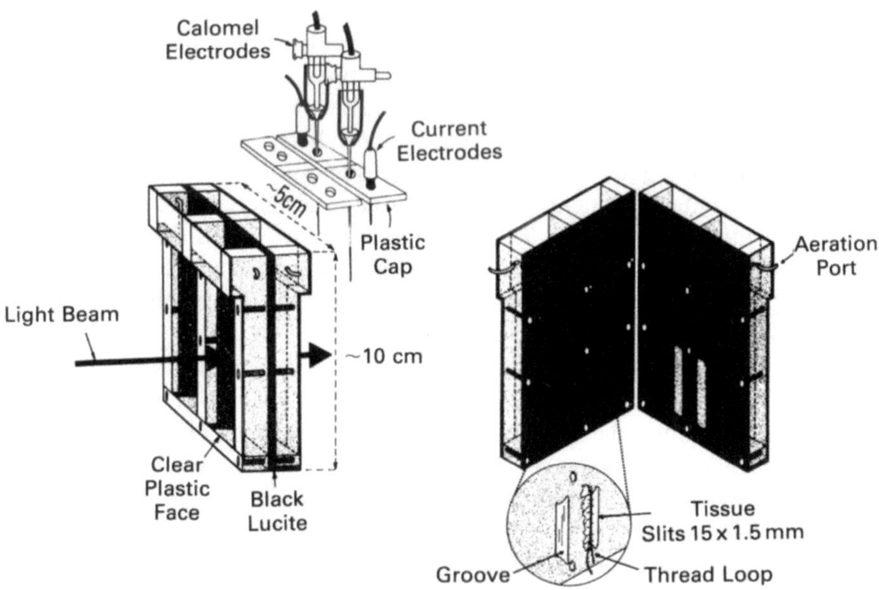

Fig. 6.5. Chamber used to perform simultaneous transport and spectrophotometric studies in epithelia using split-beam spectrophotometer. Insert: procedure used to fasten tissue to chamber. [Adapted from Mandel LJ, Moffett DF, Jöbsis FF (1975) Redox state of respiratory chain enzymes and potassium transport in silkworm midgut. Biochim Biophys Acta **408**:123–134]

requires two identical samples of tissue. It is essentially a flattened version of an Ussing double chamber with clear windows to permit spectrophotometric measurements. Each pair of half-chambers is connected by a slit measuring 15 × 1.5 mm. Fastening of the epithelia over the slits separating the half-chambers is facilitated by a grooved ridge surrounding each slit. The tissues are tied down over the slits with thread loops that fit into the grooves, making a tight seal with little edge damage. Each of the four half-chambers contains 15 ml Ringer's solution, stirred and oxygenated by vigorous bubbling with 100% oxygen, or any other appropriate gas mixture. Electrical measurements are performed by the introduction of agar bridges through a plastic cap on top of the chamber, connected to appropriate electrodes. The potential difference across each tissue is measured with calomel electrodes and current is passed through the tissue via Ag–AgCl electrodes. An automatic voltage clamp that compensates for the resistance of the solution between the potential difference bridges is used to pass the appropriate current through the tissue to maintain short-circuit conditions. Concurrently, the light beams are passed through the chamber for the spectrophotometric measurements, as shown in Fig. 6.5.

A chamber designed for use with a rapid-scanning spectrophotometer (Mandel and Riddle 1979) is shown in Fig. 6.6. This chamber requires only a single tissue sample mounted on a circular aperture either between Sylgard washers or on a groove to minimize edge damage. The aperture area can be varied between 1.65 and 0.33 cm^2 to accommodate a variety of tissue sizes. For the midgut of *Manduca sexta* (tobacco hornworm), fastening of the tissue between the half-chambers is facilitated by initially mounting the tissue on a plastic disk with a grooved ridge surrounding the central circular opening. The tissue is tied down with a thread loop, as discussed earlier. The disk with the tissue is then placed in a recessed space flush against one of the half-chambers; a slight amount of grease is used to obtain a good seal between the plastic disk and the plastic chamber. Each half-chamber contains solution stirred and oxygenated by vigorous bubbling with 100% oxygen. The potential difference across the tissue is measured and short-circuit conditions are achieved as described above. Concurrently, the light beam is passed through the chamber for the spectrophotometric measurements.

Two types of these chambers have been used in conjunction with the rapid-scanning spectrophotometer and they differ mainly in the volume of the compartments containing the bathing solutions, namely 20 and 1 ml, respectively. The chamber with the larger compartments allows electrical measurements to be made and also permits introduction of a pH electrode through the top of the secretory side to allow measurement of the acid secretory rate in gastric mucosa through the pH-stat method. The smaller chamber permits electrical measurements on the tissue but does not have sufficient space for a pH electrode. This chamber was designed for rapid changes in bathing solution to be used in kinetic experiments and rapid disassembly for tissue sampling.

The chamber shown in Fig. 6.7 was designed to perform optical studies in

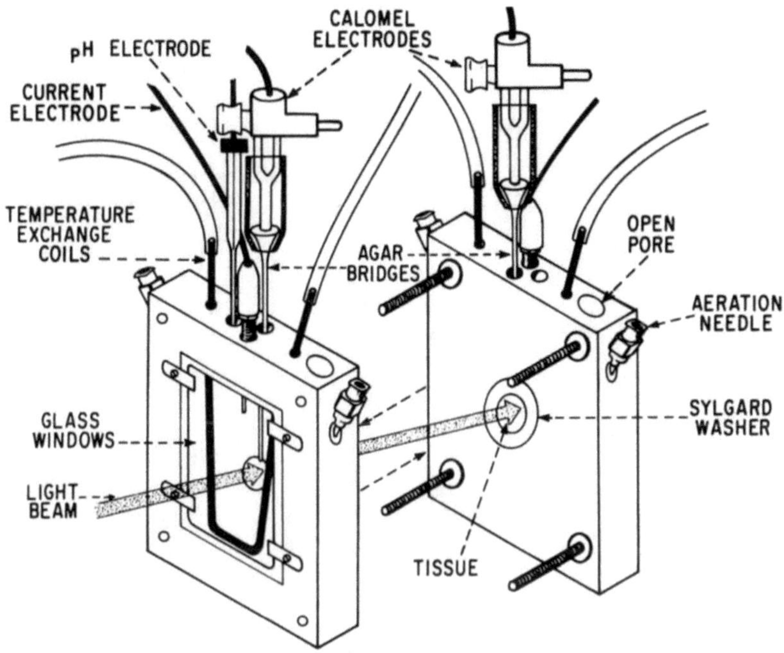

Fig. 6.6. Large version of chamber used to perform simultaneous transport and spectrophotometric studies in epithelia using rapid-scanning spectrophotometer. Procedure used for fastening tissue is identical to that shown in Fig. 6.5, except that the opening is round instead of a slit. [From Mandel LJ, Riddle TG (1979) Kinetic relationship between energy production and consumption in frog gastric mucosa. Am J Physiol **236**:E301–E308]

conjunction with oxygen consumption measurements and tissue sampling for biochemical analysis on a suspension of either kidney tubules or isolated cells (Balaban et al. 1980b). It was designed around a flat-bottomed vertical glass cylinder, in which ports were added for a Clark oxygen electrode, a diaphragm for chemical additions, and a narrow spout for sampling the suspension. The temperature is controlled at 37°C by a removable metal water jacket. To initiate an experiment, the tubule suspension is preequilibrated for 45–60 min with 75% O_2, 20% N_2, 5% CO_2 gas at 37°C. A 10-ml aliquot of this suspension is then placed into the glass chamber, which is subsequently sealed by inserting a Teflon plunger into its open top. Spectrophotometry or fluorometry of the suspension is performed by transmitting light through the chamber using light-fiber bundles, which are placed in opposing holes in the metal water jacket. Precise samples of the suspension are made by advancing the Teflon plunger with an automatic syringe dispenser (Hamilton PB 600). This method of sampling causes no dilutional or movement artifacts in either the oxygen consumption or optical measurements. Excellent mixing is maintained by a magnetic stirring bar at the

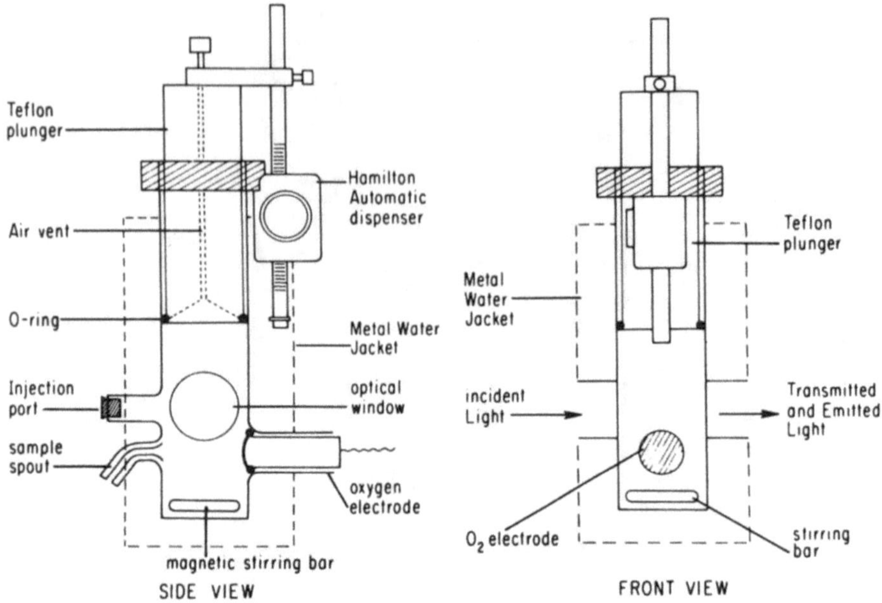

Fig. 6.7. Chamber used for simultaneous monitoring of oxygen consumption and optical properties of cell suspension. [From Balaban RS, Soltoff S, Storey JM, Mandel LJ (1980) Improved renal cortical tubule suspension: Spectrophotometric study of O_2 delivery. Am J Physiol **238**:F50–F59]

bottom of the cylinder. A slight modification of this chamber was obtained by replacing the Teflon plunger with a potassium-selective electrode and adding an opening for a reference electrode. This configuration permits the measurement of the extracellular potassium concentration in addition to the aforementioned variables.

IV. Uses of Fluorometry in Epithelia

1. Introduction

Fluorometry has been used on a number of intact tissues to monitor the redox state of mitochondrial NAD. Using this technique, it has been determined that mitochondrial state transitions similar to those described by Chance and Williams (1956) do occur in these tissues. These responses were first demonstrated in nonepithelial tissues such as muscle (Jöbsis and Duffield 1967) and brain (Rosenthal and Jöbsis 1971), in which stimulation of work caused the oxidation of mitochondrial NAD, indicating that a mitochondrial transition to a more active state occurred. The use of fluorometry in epithelial tissues was recently reviewed by Mandel and Balaban (1981) and thus is not reevaluated here. Fluorometry has not been used extensively in insect epithelia. We have used it to demonstrate that a mitochondrial uncoupler causes the oxidation of NADH in the midgut of the tobacco hornworm (Mandel et al. 1980a). Otherwise, examples for the measurement of NADH fluorescence in epithelia are more readily available in the kidney literature. Experiments performed in my laboratory on a tubule suspension of cortical tubules from the rabbit kidney (Balaban et al. 1980a) provide examples of mitochondrial state transitions as a function of cellular work, availability of metabolic substrates, and anoxia.

Prior to a presentation of these examples, it should be noted that in intact tissues it is important to examine the extent to which changes in 450-nm fluorescence reflect changes in the redox state of mitochondrial NADH. Interfering signals could originate from changes in cytoplasmic NADH or any NADPH in the tissue because they also fluoresce at 450 nm. The degree of interference needs to be evaluated individually in each tissue. Various methods have been used to determine that virtually all the measured changes in 450-nm fluorescence originate from mitochondrial NADH in skeletal muscle (Jöbsis and Duffield 1967), heart muscle (Chapman 1972), brain (O'Connor 1976), and kidney cortex (Balaban et al. 1983).

2. Changes in Cellular Work and Anoxia

The predominant work of the kidney cortex is the reabsorption of sodium (Cohen and Kamm 1976), a process that is apparently mediated by the Na,K-ATPase located predominantly at the basolateral membrane (for review see Kinne 1979). Thus, in the present experiments, alterations in cellular work were obtained by either inhibition or stimulation of Na,K-ATPase turnover. Inhibition of Na,K-ATPase turnover was obtained with ouabain. The effect of ouabain on NADH fluorescence, oxygen consumption, and ATP/ADP ratio in a suspension of cortical tubules is shown in Fig. 6.8. Oxygen consumption was inhibited about 50% and was accompanied by increased fluorescence, signifying net reduction of NAD and an increase of ATP/ADP ratio. All these changes are consistent with the occurrence of a mitochondrial transition to a less active state as a result of the decreased utilization of ATP by the Na,K-ATPase and a consequent decrease in cellular ADP levels.

The opposite transition was elicited by reintroduction of potassium to a tubule suspension bathed in a potassium-free medium, as shown in Fig. 6.9. Addition of potassium caused the rapid acceleration of oxygen consumption as the potassium was actively transported into the cells. This is not shown in Fig. 6.9 but has been measured with a potassium electrode in the medium (Harris et al. 1980). Potassium addition caused the net oxidation of NAD and a decrease in the ATP/ADP ratio, as expected for a mitochondrial transition to a more active state (Balaban et al. 1980a). These mitochondrial transitions in the intact tubule strongly suggest that the intracellular nucleotides or the ATP/ADP ratio are part of the feedback signal linking active transport and oxidative metabolism (Balaban et al. 1980a).

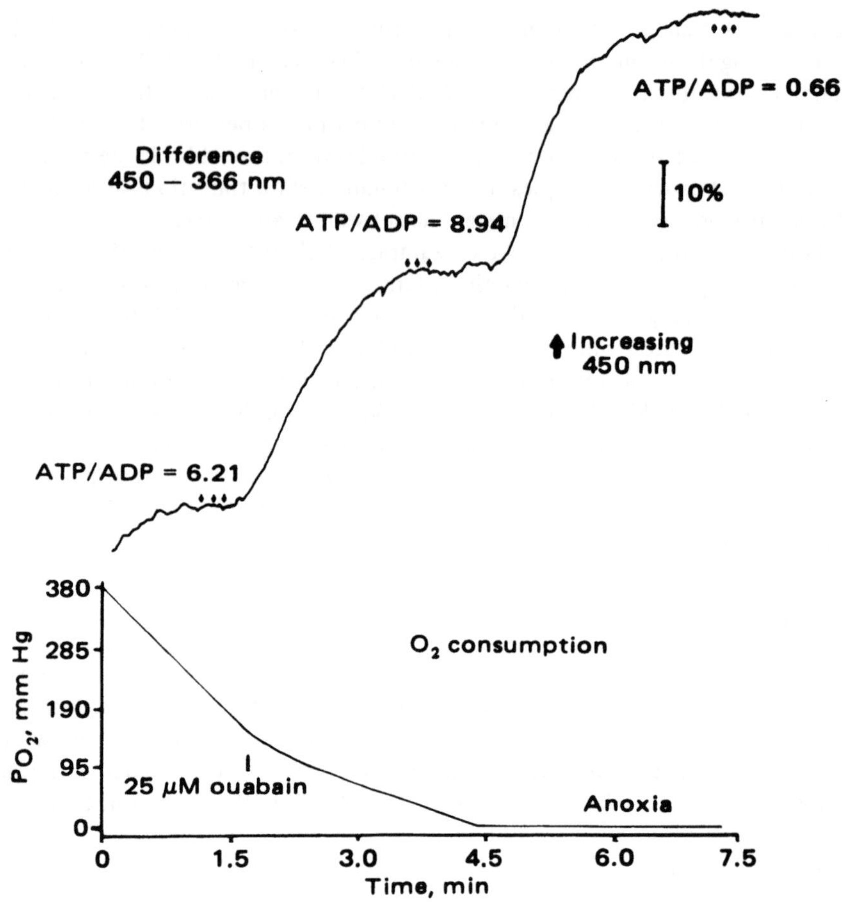

Fig. 6.8. Effect of ouabain on NADH fluorescence, oxygen consumption, and ATP/ADP ratio of cortical tubule suspension from rabbit kidney at 37°C. [From Balaban RS, Mandel LJ, Soltoff S, Storey JM (1980) Coupling of Na-K-ATPase activity to aerobic respiratory rate in isolated cortical tubules from the rabbit kidney. Proc Natl Acad Sci USA 77:447–451]

The transition to anoxia is also seen in Fig. 6.8. When all the oxygen in the chamber is consumed by the tubules, there is a rapid decrease in the ATP/ADP ratio, accompanied by the complete reduction of NADH—as would be expected for a transition to state 5 (Table 6.1).

Although the results shown in Figs. 6.8 and 6.9 are qualitatively valuable, they do not provide quantitative information regarding the degree to which these cellular respiratory conditions correspond to the classical state 4 to state 3 (and vice versa) transitions observed in isolated mitochondria. Such quantitation was recently obtained in my laboratory by Harris et al. (1981), who demonstrated that mitochondrial state 3 could be achieved in the tubule suspension by

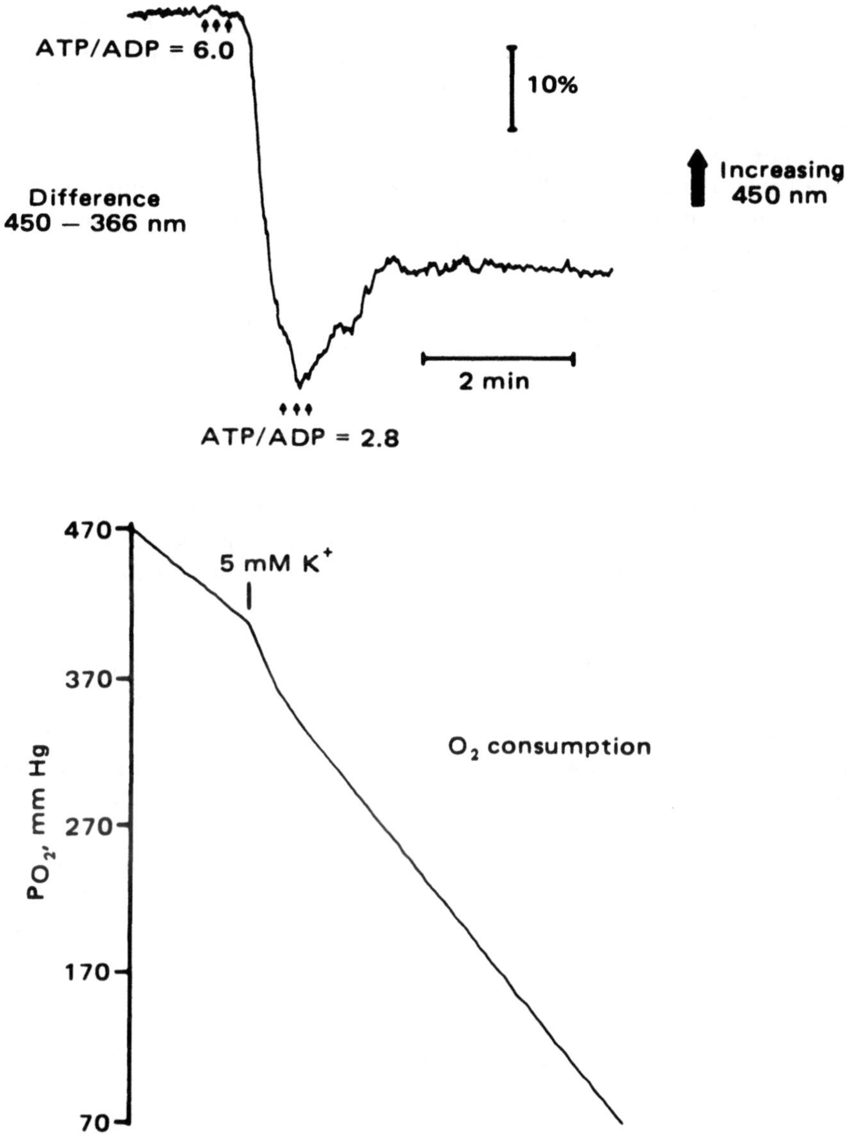

Fig. 6.9. Effect of reintroduction of potassium to a suspension of rabbit proximal tubules bathed in potassium-free medium at 37°C. Measured variables were NADH fluorescence, oxygen consumption, and the ATP/ADP ratio. [From Balaban RS, Mandel LJ, Soltoff S, Storey JM (1980) Coupling of Na-K-ATPase activity to aerobic respiratory rate in isolated cortical tubules from the rabbit kidney. Proc Natl Acad Sci USA 77:447–451]

the use of nystatin, a polyene antibiotic. This substance causes a large increase in cationic permeability, such that sodium diffuses into and potassium diffuses out of the cells. The increased intracellular sodium concentration stimulates Na,K-ATPase activity and respiration by more than 100%. By comparing this stimulated rate to that of an ADP-stimulated mitochondrial preparation from the same tissue, Harris et al. (1981) found that nystatin increased respiration in the tubular mitochondria to the state 3 rate; that is, the maximal coupled respiratory capacity in the tubules could be reached by stimulation of Na,K-ATPase activity. Having measured this maximal respiratory rate, they determined that the tissue normally operates within 50–60% of respiratory capacity. Simultaneous fluorescence studies were attempted but, unfortunately, these did not succeed because of optical interference from nystatin.

3. Detection of Substrate Availability

Another use of NADH fluorescence is to determine the metabolic substrate specificity of a tissue. When starting with a substrate-deprived condition (close to state 2), metabolic substrates may be added while NADH fluorescence is measured. The substrates' ability to supply reducing equivalents to the respiratory chain will be roughly correlated with their capacity to reduce NAD (Balaban and Mandel 1980). This type of experiment is shown in Fig. 6.10; in this experiment, performed in collaboration with Balaban, NADH fluorescence and oxygen consumption were simultaneously measured in a cortical tubule suspension. The tubules were initially exposed only to glucose as a metabolic substrate, which various investigators (Schmidt and Guder 1976), including ourselves (Balaban and Mandel 1980; Harris et al. 1981), have shown to be a relatively poor substrate in rabbit proximal tubules. A small amount of the fatty acid heptanoate was added at the time indicated, eliciting a large reduction in NAD and a stimulation of oxygen consumption; both returned to baseline levels after the added heptanoate was metabolized.

Much information may be derived from this type of experiment. First, the extent of NADH reduction and increase in oxygen consumption provide a good measure of the relative effectiveness of each metabolic substrate. We have performed dose-response experiments on a variety of substrates and obtained kinetic information (Balaban and Mandel 1980). For example, fatty acids are clearly the preferred substrates in the kidney cortex, and a half-maximal effect is obtained in the 10 μM range. Second, such a transient may be used as a measure of the rate of substrate influx, because its rate of metabolism must be a minimal measure of its rate of influx into the tissue.

These examples of the various mitochondrial state transitions in the intact tissue make it easy to appreciate how changes in NAD redox state may be used to differentiate among possible transitions. For example, in the case of a hormone or any other agent that causes simultaneous alterations in epithelial transport and metabolism, changes in NAD redox state could be used to differentiate

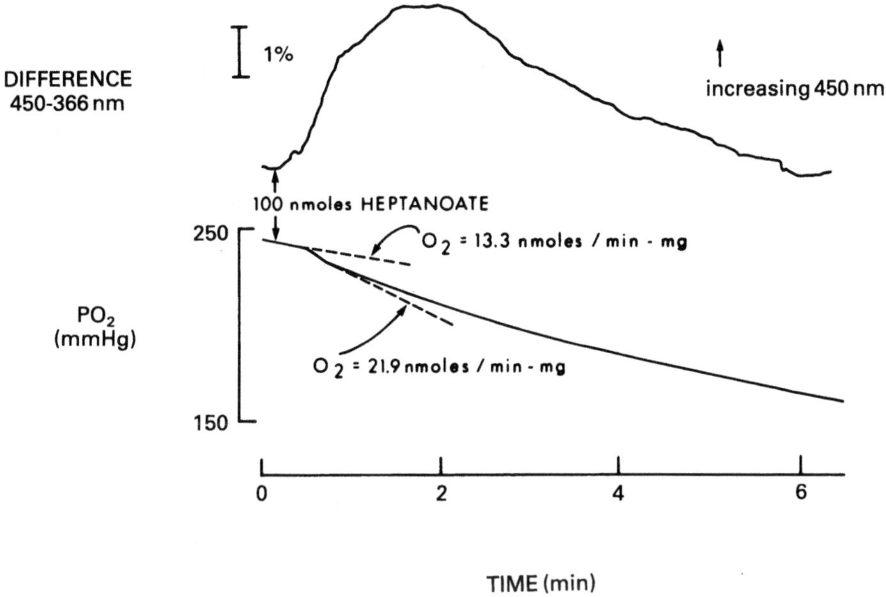

Fig. 6.10. Effect of heptanoate addition on NADH fluorescence and oxygen consumption is a substrate-deprived tubule suspension from rabbit kidney at 37°C. [From Mandel LJ (1982) Use of noninvasive fluorometry and spectrophotometry to study epithelial metabolism and transport. Fed Proc **41**:36–41]

between a primary metabolic and a primary transport event. As shown in the example in Fig. 6.9, an increase in active transport causes an increased utilization of energy, which is accompanied by an oxidation of NADH. On the other hand, increased mobilization and/or availability of metabolic substrates causes a reduction of NADH (Fig. 6.10). Therefore, appropriate experiments could differentiate between these two possibilities.

V. Uses of Spectrophotometry in Epithelia

1. Introduction

The epithelia on which spectrophotometry has been used most extensively are the frog gastric mucosa, the midguts of the silkworm and tobacco hornworm, and the proximal tubule of the rabbit kidney, although some work has also been performed on the avian salt gland. All of these tissues have one property in common, namely, a relatively high concentration of mitochondria that make the measurement of cytochrome redox changes possible. Spectrophotometry has been used to compare the relative cytochrome content of intact epithelia to that of isolated mitochondria. The approach discussed below enables the quantitative comparison between properties of isolated mitochondria and mitochondria within the intact tissue. Spectrophotometry of cytochrome a, a_3 has been used to monitor tissue anoxia as well as the kinetics leading to anoxia of the oxidative metabolic apparatus. In addition, changes in cytochrome redox state with alterations in tissue function have been used to make numerous inferences regarding the relationship between transport and metabolism in various epithelia. All of these applications of spectrophotometry in epithelia are reviewed next.

2. Cytochrome Content of Intact Epithelia

The reduced-minus-oxidized spectrum of the frog gastric mucosa appears to be very similar to that of isolated mitochondria (Hersey 1974). Analysis of the relative heights of the absorption peaks led Hersey (1974) to conclude that all the cytochromes seem to be localized within the mitochondria, as there was no evidence for the presence of extramitochondrial cytochromes. In contrast, the midgut of the silkworm shows a large shoulder at 557 nm (Fig. 6-11). Shappirio and Williams (1957) separated mitochondrial and microsomal fractions in this tissue to find a normal mitochondrial absorption spectrum and a microsomal spectrum containing only the 557-nm peak. These results, as well as kinetic studies by Chance and Pappenheimer (1954), led these investigators to conclude that the 557-nm peak corresponded to cytochrome b_5. The function of this cytochrome remains unknown, although at one point it was linked to a possible redox pump (see Sect. V.4).

Quantitation of the total amount or concentration of cytochromes in intact epithelia has been difficult because the exact path length of the light beam is unknown. The variable thickness of the tissue and the large amount of light scattering due to inhomogeneities make this a difficult measurement. One approach to this problem has been that of Harris et al. (1981). Reduced-minus-oxidized spectra of rabbit proximal renal tubules and mitochondria isolated

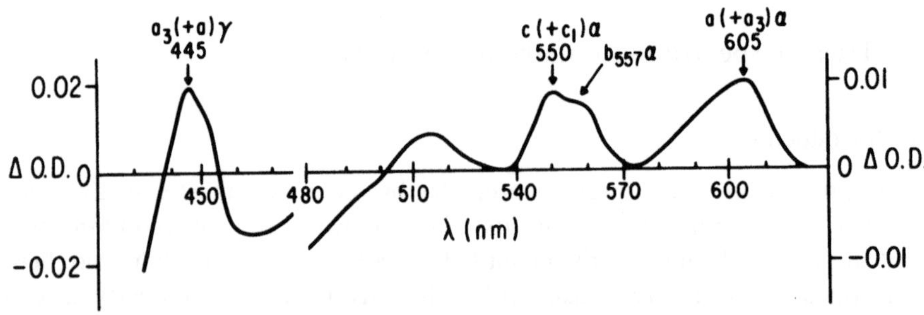

Fig. 6.11. Reduced-minus-oxidized spectrum of intact silkworm (*Hyalophora cecropia*) midgut. [From Mandel LJ, Moffett DF, Jöbsis FF (1975) Redox state of respiratory chain enzymes and potassium transport in silkworm midgut. Biochim Biophys Acta **408**:123–134]

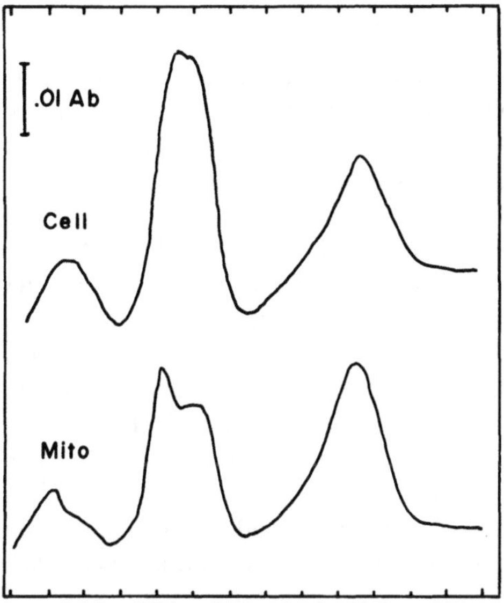

Fig. 6.12. Reduced-minus-oxidized spectra of a suspension of rabbit cortical renal tubules (Cell) and a suspension of mitochondria isolated from the rabbit cortex (Mito). Both spectra were obtained by solubilizing the suspension with Triton X-100 to minimize light scattering. [From Harris SI, Balaban RS, Barrett L, Mandel LJ (1981) Mitochondrial respiratory capacity and Na^+- and K^+-dependent adenosine triphosphatase-mediated ion transport in the intact renal cell. J Biol Chem **256**:10319–10328]

from the rabbit kidney cortex were compared in preparations solubilized with Triton X-100, which minimized scattering. As shown in Fig. 6.12, the tubule preparation shows enhanced absorption between 554 and 562 nm in comparison to the mitochondrial spectrum. Fractionation studies revealed that the additional absorption corresponds to cytochrome b_5 present in the microsomal fraction.

The amount of cytochrome a, a_3 present in each of these solubilized preparations was calculated using the extinction coefficient of 12.0 mM^{-1} cm^{-1} for the 605–630 nm wavelength pair. Harris et al. (1981) found 0.08 nmol/mg cellular protein of cytochrome a, a_3 in the tubules and 0.33 nmol/mg mitochondrial protein of that cytochrome in the mitochondrial suspension. Since all the cellular cytochrome a, a_3 is mitochondrial, these values may be used to calculate that 24% of the cellular protein is mitochondrial. In addition, knowledge of the cytochrome a, a_3 content per milligram cellular protein permits the calculation of electron turnover rate through this cytochrome from the oxygen consumption rate. Such quantitation is crucial for a direct comparison between mitochondrial function in the intact tissue and in the isolated organelle (Harris et al. 1981).

3. Redox Level of Cytochrome a, a_3 as a Measure of Tissue Anoxia

As described earlier, cytochrome a, a_3 displays an affinity for molecular oxygen in the submicromolar range, making the redox level of this cytochrome an excellent probe for tissue anoxia. An excellent example for the use of this property is a comparison of the oxygenation levels of a cortical slice from the rabbit kidney and a rabbit cortical tubule suspension.

In Fig. 6.13 results obtained in the cortical tubule suspension and cortical slices at 37°C are compared. When these tubules are prepared with open lumens and are placed in a chamber with good stirring, cytochrome a, a_3 is 94% oxidized (Balaban et al. 1980b). This redox level is independent of bath PO_2 down to very low levels, when this cytochrome becomes completely reduced. Although this experiment does not permit resolution of the PO_2 at which 50% reduction is obtained, the redox level of cytochrome a, a_3 displays a dependence on PO_2 in the tubule suspension similar to that in isolated mitochondria. This demonstrates that the tubules are well oxygenated and that oxygen diffusion is not normally limiting in the function of the tissue. In contrast, a cortical slice, even one 0.25

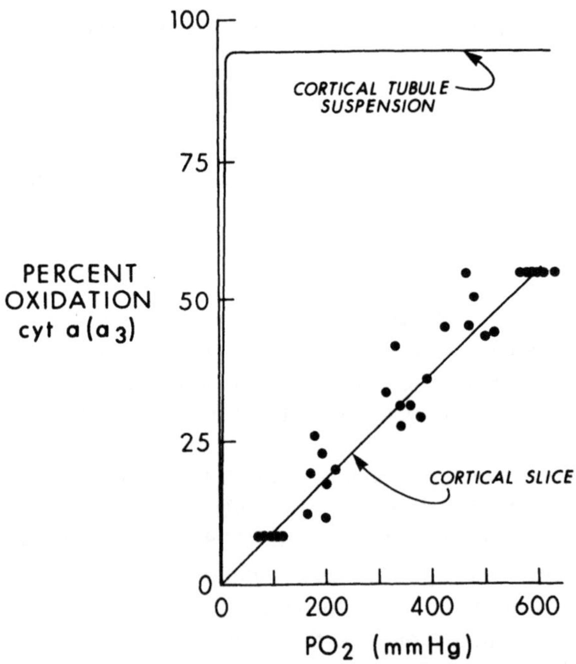

Fig. 6.13. Redox state of cytochrome a,a_3 as a function of bathing medium PO_2 in a suspension of cortical tubules compared to that in a cortical slice from rabbit kidney. [Adapted from Balaban RS, Soltoff S, Storey JM, Mandel LJ (1980) Improved renal cortical tubule suspension: Spectrophotometric study of O_2 delivery. Am J Physiol **238**:F50–F59]

mm thick, behaves very differently. The redox level of cytochrome *a*, a_3 is seen to change linearly with PO_2 and this cytochrome is only 55% oxidized at 550 mm Hg. The fact that 45% is reduced indicates that about one-half the tissue has an intracellular oxygen level below 10^{-7} *M*, which is clearly anoxic, despite the presence of high bath PO_2. Oxygen diffusion must be limiting and the size of the anoxic core appears to change as the PO_2 of the bathing medium is varied. The comparison shown in Fig. 6.13 emphasizes the importance of maintaining tissue oxygenation for normal function and provides a rather simple optical test to determine noninvasively whether a tissue is properly oxygenated.

Such a result with cortical slices brings into question the ability to properly oxygenate sheet epithelia in a chamber. The frog gastric mucosa (Hersey and Jöbsis 1969) and the midguts of the silkworm (Mandel et al. 1975) and tobacco hornworm (Mandel et al. 1980a) show 20-30% reduction of cytochrome *a*, a_3 even when their bathing media are bubbled with 100% oxygen. It is interesting that *in situ* experiments with various tissues such as rat brain (Jöbsis et al. 1977) and rat kidney (Balaban and Sylvia 1981) reveal the same level of partial reduction of cytochrome *a*, a_3 under normal conditions. One possible explanation for this behavior is that the affinity of this cytochrome for oxygen is much decreased in the intact tissue, as proposed by Jöbsis and co-workers (1977). Another possibility is that some normally functioning tissues display areas of anoxia; these areas could be continually shifting on a microscopic scale so as to represent an average of 20-30% of the tissue. Since the results shown in Fig. 6.13 demonstrate that cytochrome *a*, a_3 functions in kidney tubules as it does in isolated mitochondria, it may be concluded that the observed reduction of this cytochrome in other kidney preparations is due to areas of local anoxia.

4. Cytochrome Redox Changes and Tissue Function

The earliest report concerning the use of spectrophotometry in epithelia is by van Rossum (1968). Using avian salt gland slices, van Rossum measured NADH fluorescence and also redox levels of the mitochondrial cytochromes as active transport was altered. Inhibition of transport with ouabain caused a reduction of NAD, as expected, but oxidized cytochromes *a* and *c*, which was unexpected for a state 3 to 4 transition. However, these results can be explained if it is assumed that the slices were partially anoxic, as discussed earlier. The addition of ouabain would decrease the oxygen consumption rate, thereby increasing the depth of oxygenation; this would result in the overall oxidation of the mitochondrial cytochromes, which are very sensitive to anoxia. The response of NAD would then be a combination of the reduction caused by ouabain and the oxidation caused by the added oxygenated areas, resulting in a net reduction. Another interesting result from these studies in avian salt gland was the reduction of all respiratory carriers produced by methacholine, a stimulant of active transport in this tissue. Van Rossum (1968) concluded that methacholine, in addition to stimulating transport, also stimulated a reaction leading directly to the reduction of NAD, such as reverse electron flow. However, tissue anoxia induced by the increase in respiration by methacholine could also explain these results.

Interest in monitoring cytochrome redox states in the frog gastric mucosa and the silkworm midgut was prompted by the possibility that active transport was linked to a redox pump, deriving its energy directly from electron flow rather than a high-energy biochemical intermediate such as ATP (Hersey 1974; Mandel et al. 1975). At that time, a potassium-sensitive ATPase had been identified in the gastric mucosa (Ganser and Forte 1973) but its role in acid secretion had not been identified. Similarly, in the lepidopteran midguts, no K-ATPase nor Na,K-ATPase had been isolated (Jungreis and Vaughan 1977) that could power active potassium transport in these tissues. Given the close structural apposition between the mitochondrial and secretory membranes in these tissues (Anderson and Harvey 1966; Ito 1967) and their strong dependence on oxidative metabolism, it was reasonable to explore the possibility that redox pumps were the direct energy sources for these active transport processes. Although these studies found indirect evidence favoring the redox pump idea, more recent work in rabbit gastric mucosal glands (Malinowska et al. 1981) and the midgut of the tobacco hornworm (Mandel et al. 1980b) strongly suggests that ATP is the immediate source of energy for active transport in these tissues. Nevertheless, much information has been gained by the use of spectrophotometry in these epithelia.

In the frog gastric mucosa, alterations in acid secretory activity created unexpected changes in the mitochondrial redox state in this tissue. When the gastric mucosa was stimulated to secrete acid with a variety of secretagogues, it caused a reduction of the entire cytochrome chain and an increase in oxygen consumption (Hersey and Jöbsis 1969). Inhibition of secretion with thiocyanate caused the opposite response—an oxidation of the entire chain. These responses were contrary to what might be expected if the sole action of the secretagogues was to stimulate active transport work by the tissue, as described for the kidney.

The observed redox responses in the gastric mucosa have been interpreted by Kidder (1970) as being due to local anoxia, by Hersey and High (1972) as reflecting the properties of a redox pump and not anoxia, and by Hersey (1974) as describing a mitochondrial transition from a substrate-depleted (state 2) to an active (state 3) state. Thus, the redox responses of the gastric mucosa are complex, suggesting that the action of the secretagogues is to initiate a series of events with multiple effects on the mitochondrial redox state.

In the midgut of the silkworm, Mandel et al. (1975) observed that cytochrome b_5 was reduced with slower kinetics than the mitochondrial cytochromes in the transition to anoxia. Since the reduction kinetics of cytochrome b_5 were similar to those of the short-circuit current (measuring active potassium transport), the authors suggested that this cytochrome could be part of a redox pump directly linking the respiratory chain with the active transport system. However, further investigation in my laboratory failed to demonstrate any direct redox link between the cytochrome chain and cytochrome b_5 in preparations of isolated mitochondria and microsomes from the midgut (unpublished observations).

In the same tissue, Mandel et al. (1975) also observed an oxidation of all the cytochromes when the potassium concentration of the bathing media was decreased from 32 to 4 m*M*. Since this maneuver inhibited active potassium transport by 60%, the accompanying redox change was the opposite of that expected from a state 3 to 4 transition. Such a result would be expected if the tissue were partially anoxic, since the decreased work might decrease the size of the anoxic core, as discussed earlier. Another possibility is that the changes in tissue structure that are present with inhibited secretion of potassium (Moffett 1979) might allow better oxygenation of the tissue by removing barriers to the diffusion of oxygen.

In the midgut of the tobacco hornworm, Mandel et al. (1980a) observed cytochrome redox changes consistent with the behavior of coupled mitochondria. It was important to establish this fact because this tissue displays no change in oxygen consumption when the rate of active potassium transport is profoundly altered by large changes in the potassium concentration of the bathing media (Harvey et al. 1967; Mandel et al. 1980a). This apparent lack of respiratory control raised the question as to the degree of coupling between oxygen consumption and oxidative phosphorylation in the mitochondria of the hornworm midgut. Simultaneous monitoring of the short-circuit current (I_{sc}) and the redox level of the respiratory chain components demonstrated the following: (1) succinate (5 m*M*) reduced all the respiratory enzymes while increasing I_{sc} by 17%; (2) sesamol (5 m*M*), a mitochondrial uncoupler, reoxidized all respiratory enzymes and inhibited I_{sc} by about 50%; (3) cyanide (1 m*M*) fully reduced the cytochromes and completely inhibited I_{sc}. These redox responses indicate that the mitochondria in this tissue are normally coupled, even if respiration is maximal and is not modulated by active transport. Mitochondria isolated from the midgut show coupling and respiratory control by ADP, appearing to behave like mitochondria from other tissues. Therefore, a cytoplasmic constraint must exist in this tissue that continually elicits an unmodulated maximal respiratory rate.

In the rabbit kidney tubule suspension, the cytochrome redox levels do not change measurably during alterations in active transport work. Inhibition of Na,K-ATPase activity with ouabain, or its stimulation by potassium addition to potassium-depleted tubules, caused *no* measurable change in cytochrome redox levels (Balaban et al. 1982). This result is the expected one in reference to the redox changes observed in a mitochondrial transition between states 4 and 3 (Table 6.1). In this transition in isolated mitochondria, the oxygen consumption increases by about 800% and the cytochrome redox levels change by varying amounts, the maximum change being 19% for cytochrome *b*. In comparison, the aforementioned alterations in Na,K-ATPase activity elicited changes in oxygen consumption of 50-100% (Balaban et al. 1980a). Therefore, the changes in redox state associated with these relatively small transitions in the tubular mitochondria would be expected to be one order of magnitude smaller than those shown in Table 6.1. Such small changes in redox states, if they occurred, could not be observed in such a highly scattering medium. Clearly, the better technique for monitoring alterations in mitochondrial redox state with tissue function involves the measurement of NADH fluorescence, as described in Sect. IV.2.

In summary, the simultaneous measurement of cytochrome redox levels and epithelial function has produced mixed results that require careful consideration. In every case, the possibility of partial anoxia needs to be considered and its implications evaluated. In the avian salt gland, the frog gastric mucosa, and the silkworm midgut the results described could be due to partial anoxia. This is also true for the tobacco hornworm, except for the oxidation elicited by uncouplers. The experiments described last with kidney tubules are sobering, since they emphasize that little if any change in cytochrome redox state ought to be expected from a partial mitochondrial transition between states 3 and 4.

5. Kinetics Leading to Anoxia of Active Epithelial Transport and Cytochrome Redox Level

A somewhat different application of epithelial spectrophotometry has been based on measurement of cytochrome a, a_3 reduction kinetics at the sudden imposition of anoxia as an index of the rate of decrease in tissue oxygen consumption during the transition to anoxia. The cytochrome reduction and oxygen consumption kinetics are closely related to each other, although a nonlinear relationship exists between them (Jöbsis 1972). Respiratory rate measures the rate of energy production for a tissue that relies mainly on oxidative metabolism; the rate of inhibition of this variable reflects the kinetics of energy production during the transition to anoxia. If the kinetics leading to anoxia of a process that consumes energy are measured simultaneously, it is possible to determine whether this process is directly powered by redox energy from the cytochrome chain or whether it uses an intermediate form of energy stored in the cells, such as ATP.

We performed such a experiment in the midgut of the tobacco hornworm (Mandel et al. 1980b), as shown in Fig. 6.14. Three variables were measured in the same tissue mounted in an Ussing chamber with optical windows: the short-circuit current, which in this tissue represents active potassium transport from hemolymph to lumen (Harvey and Zerahn 1972; Moffett 1979); the redox level of cytochrome a, a_3; and the tissue ATP levels. At time zero, the oxygenated solutions bathing the tissue were rapidly changed to ones preequilibrated with 100% nitrogen, and the transition to anoxia followed. Cytochrome a, a_3 becomes rapidly reduced with a time constant of under 1 min. The short-circuit current decays considerably more slowly, with a time constant of about 3 min, and the ATP levels seem to follow the decay kinetics of the short-circuit current. In this tissue, the short-circuit current appears to measure active potassium transport even during this transient period (Blankemeyer 1981). In this highly aerobic tissue, the delay between the inhibition of energy production and the inhibition of active transport indicates that during this interval active potassium transport

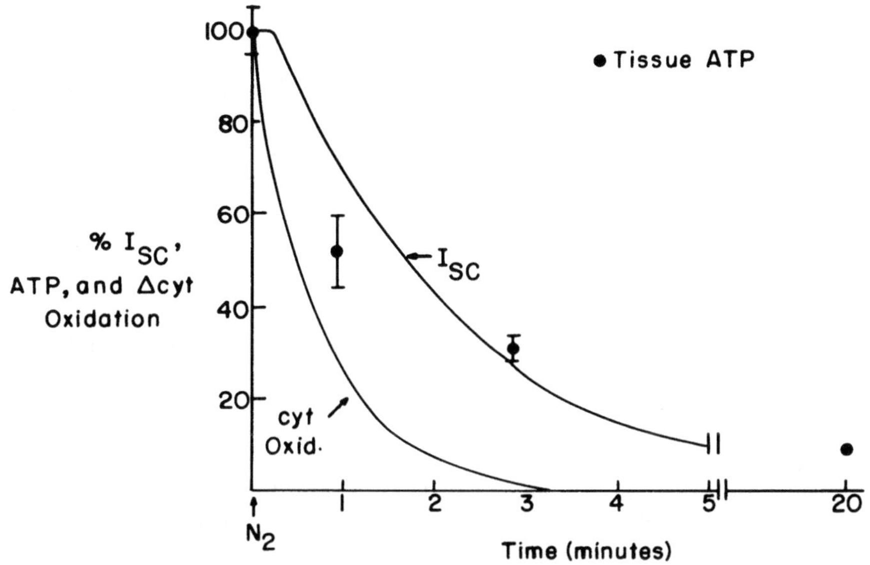

Fig. 6.14. Simultaneous plot of percentage change in short-circuit current (I_{sc}), tissue ATP, and change in cytochrome a,a_3 oxidation as a function of time in the midgut of *Manduca sexta.* At time zero, anoxic bathing solutions were introduced in the chamber. [From Mandel LJ, Riddle TG, Storey JM (1980) Role of ATP in respiratory control and active transport in tobacco hornworm midgut. Am J Physiol **238**:C10–C14]

is powered by a stored metabolic intermediate, probably ATP. This is an important finding in this tissue, inasmuch as the immediate source of energy for active potassium transport was unknown; an approach such as the present one was required to suggest that ATP could be the energy source for this process.

Wolfersberger et al. (1982) have identified and partially purified a potassium-modulated ATPase from a plasma membrane fraction of *Manduca sexta* midgut. The K_m of this ATPase is decreased about threefold by added potassium ion. At an ATP concentration of 1–3 m*M*, when the enzyme is acting at a submaximal rate, its activity can be stimulated up to 50% by added potassium. The ATP concentrations in midgut found by Mandel et al. (1980b) are precisely those required for the activity of this new ATPase to be modulated by potassium ion concentration.

VI. Concluding Remarks

Numerous examples are given in this chapter of the uses of fluorometry and spectrophotometry to measure the redox level of respiratory chain components in various epithelia. Regrettably, these optical techniques have not been extensively used in insect epithelia; therefore, most of the examples given are from work in other epithelial tissues. It is my hope that describing the various applications of this optical technology will stimulate further work in this direction in insect epithelia.

Acknowledgments. I would like to thank Dr. Mary Chamberlin for helpful critical comments on the manuscript, and Ms. Cynthia Rosenbloom for the typing.

References

Anderson E, Harvey WR (1966) Active transport by the Cecropia midgut II. Fine structure of the midgut epithelium. J Cell Biol **31**:107–134

Avi-Dor Y, Olson JM, Doherty MD, Kaplan ND (1962) Fluorescence of pyridine nucleotides in mitochondria. J Biol Chem **237**:2377–2383

Balaban RS, Mandel LJ (1980) Steady state kinetics of NAD reduction by metabolic substrates in the intact renal cell. Proceedings of the International Union of Physiological Sciences, 28th International Congress, Vol 14, p 310. Hungarian Physiological Society, Budapest

Balaban, RS, Mandel, LJ, Soltoff S, Storey JM (1980a) Coupling of Na-K-ATPase activity to aerobic respiratory rate in isolated cortical tubules from the rabbit kidney. Proc Natl Acad Sci USA **77**:447–451

Balaban RS, Soltoff S, Storey JM, Mandel LJ (1980b) Improved renal cortical tubule suspension: Spectrophotometric study of O_2 delivery. Am J Physiol **238**:F50–F59

Balaban RS, Sylvia AL (1981) Spectrophotometric monitoring of O_2 delivery to the exposed rat kidney. Am J Physiol **241**:F257–F262

Balaban RS, Dennis VW, Mandel LJ (1981) Microfluorometric monitoring of NAD redox state in isolated perfused renal tubules. Am J Physiol **240**: F337–F342

Balaban RS, Harris SI, Soltoff SP, Storey JM, Mandel LJ (1983) Cellular metabolic control. The relationship between active ion transport and aerobic metabolism in rabbit kidney cortex. Submitted for publication

Blankemeyer JT (1981) Characteristics of the decay of the spontaneous potential difference due to active potassium transport in the insect midgut epithelium. Fed Proc **40**:374 (Abstract)

Chance B (1951) Rapid and sensitive spectrophotometry III. A double-beam apparatus. Rev Sci Instrum **22**:634–638

Chance B (1954) Spectrophotometry of intracellular respiratory pigments. Science **120**:767–775

Chance B (1957) Techniques for the assay of the respiratory enzymes. In: Colowick SP, Kaplan NO (eds) Methods in enzymology, Vol IV, pp 273–329. Academic Press, New York

Chance B, Jöbsis FF (1959) Changes in fluorescence in a frog sartorius muscle following a twitch. Nature **184**:195–196

Chance B, Pappenheimer AM (1954) Kinetic and spectrophotometric studies of cytochrome b_5 in midgut homogenates of Cecropia. J Biol Chem **209**:931–943

Chance B, Williams CM (1956) The respiratory chain and oxidative phosphorylation. Adv Enzymol **17**:65–134

Chance B, Cohen P, Jöbsis FF, Schoener B (1962) Localized fluorometry of oxidation-reduction states of intracellular pyridine nucleotide in brain and kidney cortex of the anesthetized rat. Science **136**:325

Chapman JP (1972) Fluorometric studies of oxidative metabolism in isolated papillary muscle of the rabbit. J Gen Physiol **59**:135–154

Cohen JJ, Kamm DE (1976) Renal metabolism: Relation to renal function. In: Brenner BM, Rector FC (eds) The kidney, pp 126–200. Saunders, Philadelphia

Ganser AL, Forte JG (1973) K^+-Stimulated ATPase in purified microsomes of bullfrog oxynitic cells. Biochim Biophys Acta **307**:169–180

Hansford RG (1980) Control of mitochondrial substrate oxidation. Curr Top Bioenerg **10**:217–278

Harris SI, Balaban RS, Mandel LJ (1980) Oxygen consumption and cellular ion transport. Evidence that the ATP/O_2 ratio is near 6 in the intact cell. Science **208**:1148–1150

Harris SI, Balaban RS, Barrett L, Mandel LJ (1981) Mitochondrial respiratory capacity and Na^+- and K^+-dependent adenosine triphosphatase-mediated ion transport in the intact renal cell. J Biol Chem **256**:10319–10328

Harvey WR, Zerahn K (1972) Active transport of potassium and other alkali metals by the isolated midgut of the silkworm. Curr Top Membr Transp **3**: 367–410

Harvey WR, Haskell JA, Zerahn K (1967) Active transport of potassium and oxygen consumption in the isolated midgut of Hyalophora cecropia. J Exp Biol **46**:235–248

Hersey SJ (1974) Interactions between oxidative metabolism and acid secretion in gastric mucosa. Biochim Biophys Acta **344**:157–203

Hersey SJ, High WL (1972) Effect of unstirred layers on oxygenation of frog gastric mucosa. Am J Physiol **223**:903–909

Hersey SJ, Jöbsis FF (1969) Redox changes in the respiratory chain related to acid secretion by the intact gastric mucosa. Biochem Biophys Res Commun **36**:243–250

Ito S (1967) Anatomic structure of the gastric mucosa. In: Handbook of physiology, Sect 6, Vol II, pp. 705–741. American Physiological Society, Washington, DC

Jöbsis FF (1972) Oxidative metabolism at low PO_2. Fed Proc **31**:1404–1413

Jöbsis FF, Duffield JC (1967) Oxidative and glycolytic recovery metabolism in muscle. J Gen Physiol **50**:1009–1047

Jöbsis FF, Keizer HJ, LaManna JC, Rosenthal M (1977) Reflectance spectrophotometry of cytochrome aa_3 in vivo. J App Physiol **43**:858–872

Jöbsis FF, Stainsby WN (1968) Oxidative of NADH during contractions of circulated mammalian skeletal muscle. Respir Physiol **4**:292–300

Joyner RW, Moore JW (1973) Computer Controlled Voltage Clamp Experiments. Ann Biomed Eng **1**:368–380

Jungreis AM, Vaughan GL (1977) Insensitivity of lepidopteran tissues to ouabain: absence of ouabain binding and Na^+/K^+-ATPase in larval and adult midgut. J Insect Physiol **23**:503–509

Kidder III GW (1970) Unstirred layers in tissue respiration: Application to studies of frog gastric mucosa. Am J Physiol **219**:1789–1795

Kinne R (1979) Metabolic correlates of tubular transport. In: Giebisch G, Tosteson DC, Ussing HH (eds) Membrane transport in biology. Springer-Verlag, Berlin, **4B**:529–562.

Malinowska DH, Koelz HR, Hersey SJ, Sachs G (1981) Properties of the gastric proton pump in unstimulated permeable gastric glands. Proc Natl Acad Sci USA **78**:5908–5912

Mandel LJ, Balaban RS (1981) Stoichiometry and coupling of active transport to oxidative metabolism in epithelial tissues. Am J Physiol **240**:F357–F371

Mandel LJ, Riddle TG (1979) Kinetic relationship between energy production and consumption in frog gastric mucosa. Am J Physiol **236**:E301–E308

Mandel LJ, Moffett DF, Jöbsis FF (1975) Redox state of respiratory chain enzymes and potassium transport in silkworm midgut. Biochim Biophys Acta **408**:123–134

Mandel LJ, Riddle TG, LaManna JC (1976) A rapid scanning spectrophotometer and fluorometer for in vivo monitoring of steady-state and kinetic optical properties of respiratory enzymes. In: Jöbsis FF (ed) Oxygen and physiological function, pp 79–89. Professional Information Library, Dallas

Mandel LJ, Moffett DF, Riddle TG, Grafton MM (1980a) Coupling between oxidative metabolism and active transport in the midgut of the tobacco hornworm. Am J Physiol **238**:C1–C9

Mandel LJ, Riddle TG, Storey JM (1980b) Role of ATP in respiratory control and active transport in tobacco hornworm midgut. Am J Physiol **238**:C10–C14

Moffett DW (1979) Bathing solution tonicity and potassium transport by the midgut of the tobacco hornworm Manduca sexta. J Exp Biol **78**:213–223

O'Connor MJ (1976) Origin of labile NADH tissue fluorescence. In: Jöbsis FF (ed) Oxygen and physiological function, pp 90–99. Professional Information Library, Dallas

Oshino N, Sugano T, Oshino R, Chance B (1974) Mitochondrial function under hypoxic conditions: The steady states of cytochrome a + a_3 and their relation to mitochondrial energy states. Biochim Biophys Acta **368**:298–310

Rosenthal M, Jöbsis FF (1971) Intracellular redox changes in functioning cerebral cortex. II. Effects of direct cortical stimulation. J Neurophysiol **34**: 750–762

Schmidt U, Guder WG (1976) Sites of enzyme activity along the nephron. Kidney Int **9**:233–242

Shappirio DG, Williams CM (1957) The cytochrome system of the Cecropia silkworm II. Spectrophotometric studies of oxidative enzyme systems in the wing epithelium. Proc R Soc Lond (Biol) **147**:233–246

van Rossum GDV (1968) Relation of the oxidoreduction level of electron carriers to ion transport in slices of avian salt gland. Biochim Biophys Acta **153**:124–131

Wolfersberger MG, Harvey WR, Cioffi M (1982) Transepithelial potassium transport in insect midgut by an electrogenic alkali metal ion pump. Curr Top Membr Transp **16**:109–133

Yang CC (1954) A rapid and sensitive recording spectrophotometer for the visible and ultraviolet region. I. Description and performance. Rev. Sci Instrum 25:801–807

Yang CC, Legallais V (1954) A rapid and sensitive recording spectrophotometer for the visible and ultraviolet region. II. Electronic circuits. Rev Sci Instrum 25:807–813

Chapter 7

Coulometric Measurement of Oxygen Consumption in Insects

A. A. Heusner and M. L. Tracy

I. Introduction

The energy metabolism of an animal may be affected by environmental factors (temperature, light, partial pressure of oxygen, etc.) as well as organismic properties (reproductive and nutritional status, stage of development, etc.). Changes in energy metabolism due to specific metabolic factors can only be interpreted in a meaningful way when all other organismic and environmental factors have been standardized or controlled during an experiment. In practice, this is best achieved when the energy metabolism of an animal is measured over long periods of time under environmental conditions that approximate those of the normal habitat. Of particular importance in the interpretation of metabolic data is the establishment of normal daily changes in energy metabolism due to circadian rhythms in locomotor activity. In general, animals display their daily periodicity only in an environment that does not stress them. For example, a single honeybee survives well for more than 1 week when the conditions inside the respiratory chamber approximate those in the beehive (the presence of beeswax and honey), but would die within a few hours when deprived of those conditions (Heusner and Stussi 1964). *Drosophila* display striking circadian variations in energy metabolism when placed in a simple glass respiratory chamber that contains a glucose-agar medium (Heusner 1970).

Oxygen consumption is the best indicator of total energy metabolism in aerobic animals. It can be automatically recorded over long periods of time by using coulometry, in which the oxygen consumed by an animal is quantitatively replaced by electrolytically generated oxygen, the amount produced being calculated using Faraday's law. This technique is particularly well suited for the range of oxygen consumption observed in insects. Many respirometers have been based on this principle (Bennet-Clark 1932; Brown 1954; Capraro 1953; Chase et al. 1968; DeBoer 1929; Fernandes 1923; Fourche 1964; Hayward et al. 1963; Heusner and Jameson 1981; Heusner and Ruhland 1959; Heusner and Stussi 1964; Heusner et al. 1965; Heusner et al. 1982; Klekowski and Zajdel 1972; MacFayden 1961; Moyat 1957; Stussi 1967; Swaby and Passey 1953; Turner and Stevenson 1974; Ulrich 1940; Wager and Porter 1961; Werthessen 1937; Winteringham 1959; Woodland and Clack 1975).

In this chapter the basic principles of the coulometric measuring process are discussed. Two digital coulometric respirometers are described that permit numerical recording of oxygen consumption from a few nanoliters to 150 ml/h. The digital electronics are designed so as to permit interfacing of the recording system with computers and microprocessors. Special emphasis is placed on long-term automatic recording of oxygen consumption under sterile conditions.

II. Theory of Coulometric Measurement

The amount of oxygen consumed by an animal in a closed system can be measured directly, as a decrease in volume of the system, or indirectly, through application of the ideal gas law, as a decrease in pressure of the system. Such measurements are valid only if carbon dioxide is completely absorbed and all other factors causing changes in volume or pressure are held constant (barometric pressure, temperature, partial pressure of water). These methods would severely limit the duration of metabolic experiments due to oxygen depletion. However, if the consumed oxygen is quantitatively replaced with new oxygen, the decline in the partial pressure of oxygen is prevented. Manual replacement of very small volumes of oxygen (in the microliter range) in a precise, quantitative manner requires laborious manipulations and almost continuous attention of a technically skilled observer. These constraints limit both the frequency with which measurements can be made and the duration of experiments. Such limitations are particularly undesirable in studies of metabolic rhythms.

Electrolytic production of oxygen in a closed system permits replacement of consumed oxygen, as well as accurate and precise measurement, through the application of Faraday's law, of the amount of oxygen replaced. When this coulometric measuring process is automated, metabolic experiments can be performed over long periods of time. The automatic measuring process consists of three basic technical operations: (1) the gasometric detection of a decrease in volume or pressure in the closed system; (2) the quantitative replacement of the consumed oxygen with electrolytically produced oxygen; and (3) the coulometric determination of the amount of oxygen produced.

The two first operations maintain a constant volume and gas composition in the closed system in the face of continuous withdrawal of oxygen by the animal. The third operation is an application of Faraday's law to the electrolytic production of oxygen from a saturated $CuSO_4$ solution. The overall electrochemical process is summarized by the following reaction:

$$2\,CuSO_4 + 2\,H_2O \rightarrow 2\,H_2SO_4 + 2\,Cu + O_2. \tag{7.1}$$

In practice, oxygen is released at a platinum anode and copper is deposited at the cathode. With appropriate cathodic current density there is no formation of hydrogen at the cathode (see Sect. V). However, there is the possibility of a low oxygen uptake due to copper deposition (Bockris and Enyo 1962; Richards et al. 1900), but this source of error can be made negligible by adjusting the volume of a compensating thermobarometer. The amount of oxygen released at the anode is computed from the quantity of electricity q involved in the process:

$$V_{O_2} = \frac{V_M}{4\mathcal{F}} \cdot q \tag{7.2}$$

where V_{O_2} is the volume of oxygen produced in microliters STPD; V_M is the molar volume, 22.413×10^6 µl; $\mathcal{F}$ is 1 faraday or 96,487 coulombs (C); and q is the quantity of electricity in coulombs.

Since V_M and $\mathcal{F}$ are physical constants, Faraday's law can be rewritten

$$V_{O_2} = kq \tag{7.3}$$

where k is the coulometric constant of 58.073 µl/C. The measure of the quantity of electricity q can be based on the electrostatic or the electrodynamic definition of q, hence there are two coulometric methods:

Capacitive coulometry is based on the electrostatic expression of q (Heusner 1970; Heusner et al. 1982; Klebowski and Zajdel 1972):

$$q = C \cdot \Delta V \tag{7.4}$$

where C represents the capacitance and ΔV the decrease in voltage of the capacitor during the discharge. In this method a capacitor is discharged through a saturated $CuSO_4$ solution to produce oxygen.

Chronocoulometry is based on the electrodynamic expression of q (Heusner and Jameson 1981; Heusner et al. 1965):

$$q = I \cdot t \tag{7.5}$$

where I is the current intensity and t the time during which the current I flows. If I is electronically held constant, the volume of oxygen produced is proportional to t; hence, a volume measurement can be replaced by a time measurement that is technically easy to implement.

Note that in combining Eqs. (7.2) and (7.5) we can write

$$\dot{V}_{O_2} = 0.0581 \cdot i \tag{7.6}$$

Equation (7.6) shows that the instantaneous rate of oxygen consumption $\dot{V}_{O_2}$ (in microliters per second) is linearly related to the instantaneous current intensity i (in milliamperes). Therefore, the possibility exists, at least in theory, to record the instantaneous oxygen consumption if the rate of oxygen production could be instantaneously matched to the oxygen consumption of the animal. This has proved to be very difficult in practice because of instabilities in the negative feedback mechanism. Relatively large time constants (on the order of 5 min) would have to be introduced into the feedback loop in order to dampen or remove the oscillations, which in turn would defeat the goal of instantaneous measurement. We will not discuss implementation of Eq. (7.6) in this chapter.

III. Capacitive Coulometric Microrespirometry

1. Description of Components

Respiratory chambers of various forms and sizes can be adapted to the size of the animal to be studied. The two respiratory chambers shown in Fig. 7.1 were designed specifically for insects. Ground-glass joints are used to connect the chamber to the electrolytic cell (not shown) of the coulometric microrespirometer. The upper end of the respiratory chamber is designed according to the animal's nutritive needs. Carbon dioxide produced by the animal is absorbed by baralyme, which is held in the stem of the chamber by a small strip of Teflon. The baralyme is separated from the animal by a thin layer of cotton (Fig. 7.1a), which prevents the animal from reaching the chemical, allows rapid diffusion of gases, and keeps the inside of the respiratory chamber sterile. The bulb chamber in Fig. 7.1a (1-3 ml) has been used for metabolic studies in mosquitoes (*Aedes aegypti;* Heusner and Lavoipierre 1973; Heusner et al. 1973) and fruit flies (*Drosophila melanogaster,* Kayser and Heusner 1967). The chamber shown in Fig. 7.1b is suitable for aquatic larvae.

The glass microrespirometer is a dissymetrical differential system that consists of a small respiratory chamber, a very large thermobarometer, a manometer, and the cell of electrolysis (microanode, cathode, and $CuSO_4$ solution). The respiratory chamber shown in Fig. 7.2 is yet another design that has been used to record oxygen consumption in ticks (*Dermacentor andersoni*; Heusner et al. 1982). The manometer, which is contained in the electrolytic cell, is a glass tube, 85 mm long, with an inside diameter of 4 mm. It is centered in a glass joint that fits a cylindrical Kimax weighing bottle of 16-ml capacity containing the $CuSO_4$ solution. Its upper end is connected to the respiratory chamber by a glass joint, its lower end projecting into the $CuSO_4$ solution. The lower end of the manometer contains a platinum microanode and a Teflon ring. The platinum microanode is a platinum wire sharpened by an electrolytic process so that its tip is less than 0.1 μm in diameter (Wolbarsht et al. 1960). To make sure no current flows between the $CuSO_4$ solution and the point where the platinum microanode is sealed into the manometric tube, the microanode is inserted into the manometer 25 mm above the upper rim of the Teflon ring. The $CuSO_4$ solution

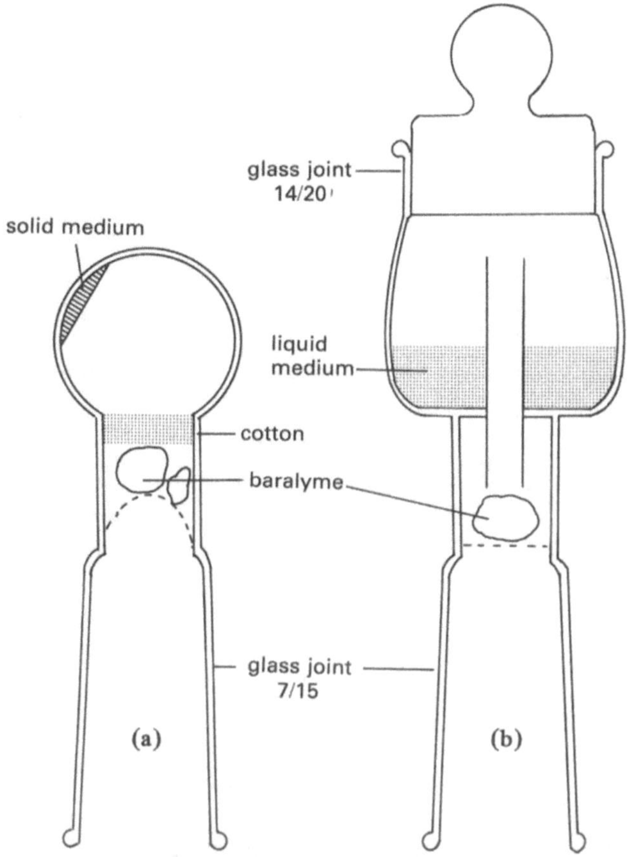

Fig. 7.1. Respiratory chambers. **(a)** Bulb chamber suitable for fruit flies and mosquitoes. **(b)** Chamber used for aquatic insects.

also serves as the manometric fluid, which separates the air space of the respiratory chamber (V_R) from that of the thermobarometer (V_T). The height of the column of $CuSO_4$ solution in the manometer is critical, since it determines the rate of oxygen diffusion between V_R and V_T. An insufficiently long column is therefore a potential source of error.

The cathode is a Teflon-insulated platinum wire, 0.4 mm in diameter, which is immersed in the $CuSO_4$ solution outside of the manometer. Its insulation is stripped from the last 5 mm; it is important to keep the insulation at the air-solution interface to prevent oxidation of copper. Both the cathode and micro-anode are connected to Teflon-insulated copper wires, which are hermetically sealed into port a (Fig. 7.1) with water-resistant epoxy resin.

A thermobarometer encloses both the respiratory chamber and the cell of electrolysis. It is a glass tube (290-ml volume) with a glass joint at the top. Its

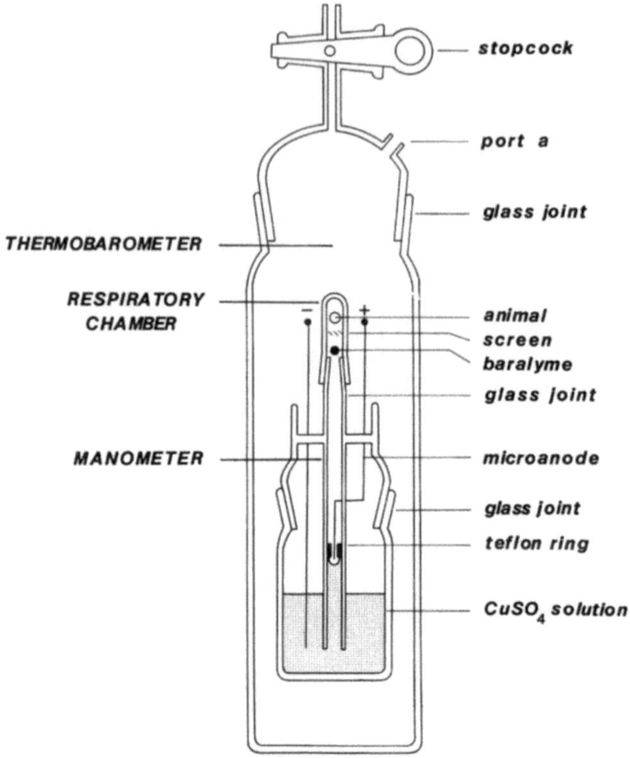

Fig. 7.2. Glass microrespirometer.

air space communicates freely with the cell of electrolysis. The dissymmetry in the volume of the respiratory chamber and the thermobarometer (1:1160) significantly increases the sensitivity of detection, while not interfering with compensation for temperature and pressure changes (Heusner 1970; Heusner et al. 1965). It also makes negligible the pressure change in the thermobarometer caused by the low oxygen uptake that is associated with the copper deposition at the cathode. The stopcock on top of the thermobarometer is necessary for the equilibration of pressure at the beginning of an experiment.

The microrespirometer functions as follows: A decrease in volume in the respiratory chamber as the animal consumes oxygen causes the $CuSO_4$ meniscus in the manometer to rise until the meniscus comes into contact with the platinum microanode. On contact, current flows through the $CuSO_4$ solution and produces oxygen. The oxygen is released into the respiratory chamber and the increasing pressure returns the meniscus to its initial form when the contact between the $CuSO_4$ solution and the platinum microanode is open. The actual change in volume or pressure in the respiratory chamber is very small: less than 1 nl or 0.001 mm Hg in a 1-ml air space. Notice that the platinum microanode serves as the detecting as well as the oxygen-producing electrode.

Since the coulometric measurement is made when the current flows through the $CuSO_4$ solution, each measurement begins with the establishment of contact between the detecting microanode and the $CuSO_4$ solution and ends with the breaking of that contact. If, after each bout of electrolysis, the air space V_R is returned to its initial volume, the amount of oxygen produced is equal to that which has been consumed by the animal. For this to be true, the meniscus must take the same form at the same level in the manometric tube at the end of each bout. Consequently, the smallest measurable volume of oxygen is that which is produced during one bout of electrolysis.

The smallest bout is limited by the amount of energy necessary to break contact between the microanode and the $CuSO_4$ solution. As the meniscus moves away, the microanode retains a film and causes the meniscus to bulge. The final detachment requires mechanical work to overcome the effect of interfacial energies between the $CuSO_4$ solution, the platinum microanode, the nascent oxygen, and the wall of the manometric tube. In the absence of any external mechanical energy, such as vibrations, this mechanical work is done solely by the concomitant changes in pressure and volume caused by the release of the nascent oxygen.

The sensitivity of the method is limited by the surface tension of the solution, the size of the tip of the microanode, the inside radius of the manometer, and the volume of the respiratory chamber. When the smallest amount of oxygen released is significantly larger than the oxygen consumed by an animal during the release, the measuring process becomes discontinous: the animal must consume the amount of oxygen released before a new measurement can be made. However, when the released oxygen is no longer the only source of mechanical energy to break contact, such as in the presence of mechanical vibrations, the sensitivity is no longer limited by the factors enumerated above and the measuring process becomes practically continuous.

Experiments have shown that detection is improved in the presence of mechanical vibrations when a Teflon ring is inserted into the manometric tube slightly above the tip of the microanode. The volume of oxygen necessary to break contact in this case drops from 10 to about 0.02 nl, thereby increasing the sensitivity by a factor of 500. The magnitude of this increase exceeds by far the effect expected from a reduction of the diameter alone. Of particular interest is the fact that the position of the ring relative to the platinum microanode is extremely critical (within less than 0.1 mm) for obtaining this effect. This phenomenon is due to the particular behavior of the meniscus at the hydophobic Teflon ring (Heusner et al. 1982).

The behavior of the meniscus when the pressure in the air space above it increases continuously is shown in Fig. 7.3. The presence of the Teflon ring forces the meniscus to change its radius of curvature when pressure inside the respiratory chamber changes. When oxygen is released into the respiratory chamber, the meniscus becomes concave and decreases its radius, while its perimeter remains stationary at the inner edge of the Teflon ring (Fig. 7.3, state 1).

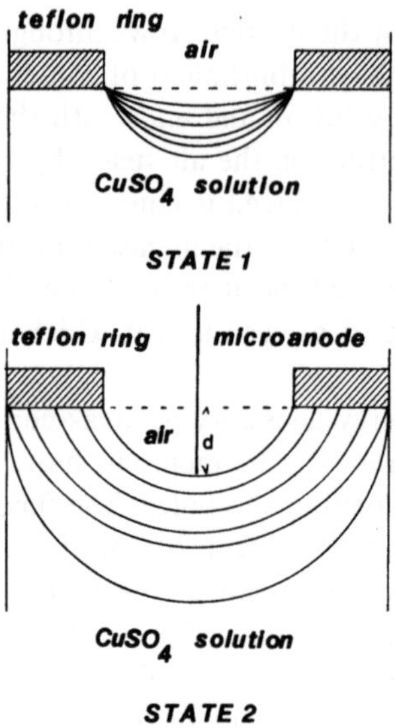

Fig. 7.3. Behavior of the $CuSO_4$ meniscus at the Teflon ring.

This deformation ends when the angle between the tangent to the meniscus and the horizontal portion of the Teflon ring reaches the contact angle of the $CuSO_4$ solution with Teflon (the contact angle between water and Teflon is 108°; Adam 1964). The meniscus is now suspended from the horizontal Teflon surface. As the pressure continues to build, the radius of the meniscus increases while its perimeter moves from the inner to the outer edge of the ring, the contact angle between the $CuSO_4$ solution and the Teflon remaining constant (Fig. 7.3, state 2). Of particular interest is the fact that in state 2, as pressure in the respiratory chamber increases, the transmeniscus pressure actually decreases when the radius of the meniscus increases (Laplace's law).

Observation of the meniscus reveals that the sensitivity of gasometric detection increases suddenly when the meniscus is suspended from the horizontal surface of the Teflon ring. Physical analysis of the sensitivity of gasometric detection (Heusner et al. 1982) shows that it is independent of the diameter of the manometer and that it is maximum when the tip of the microanode is at a distance d below the Teflon ring, given by

$$d = 0.056\,(V_{\mathrm{R}_0})^{0.25} \tag{7.7}$$

when the inside radius of the Teflon ring (r_{i}) is

$$r_{\mathrm{i}} = 0.056\left[\frac{1+\cos\theta}{\sin\theta}\right][V_{\mathrm{R}_0}]^{0.25} \tag{7.8}$$

where V_{R_0} is the volume of the respiratory chamber when the meniscus is flat, and θ is the supplement of the contact angle between the $CuSO_4$ solution and the Teflon ring (72°).

2. Procedure

The various parts of the microrespirometer may be stored at the desired temperature before being assembled to facilitate temperature equilibration prior to the actual measurement of oxygen consumption. The entire apparatus (contained within the glass thermobarometer) is then immersed in a water bath that is maintained at the desired temperature with a tolerance of ±0.01°C.

The level of $CuSO_4$ in the manometer is adjusted prior to the beginning of temperature equilibration such that the $CuSO_4$ meniscus is close to the lower edge of the Teflon ring. This procedure is accomplished by (1) adjusting the level of $CuSO_4$ in the vial in which the manometer is contained to approximately 1 cm below the lower edge of the Teflon ring, and (2) positioning the meniscus just below the lower edge of the Teflon ring. Fine adjustments may be made by injecting or withdrawing air from the system via the stopcock at the top of the thermobarometer (Fig. 7.2). Since the air space of the thermobarometer is continuous with the air space above the $CuSO_4$ solution, any change in air volume introduced through the stopcock will result in movement of the $CuSO_4$ meniscus. However, care should be taken to prevent the introduction or removal of too large a quantity of air, which would result in pressurizing or depressurizing of the system with concomitant changes in the partial pressure of gases. Gas tensions must be equilibrated across the boundary of the $CuSO_4$ meniscus before reliable recordings can be obtained. Otherwise, apparent changes in oxygen consumption will be recorded that are a function of the movement of gases from one compartment to the other (i.e., respiratory chamber to thermobarometer or vice versa) and not the result of oxygen consumption by the animal.

Once temperature equilibration has taken place between the water bath and the microrespirometer, a fine adjustment of the level of the $CuSO_4$ meniscus is made such that it hangs from the lower edge of the Teflon ring in the manometer and is in contact only with the very tip of the microanode. At this point in time, the stopcock is closed and the recording of oxygen consumption begins.

3. Electronic Circuit

The wiring diagram of the oxygen circuit is shown in Fig. 7.4. During the charge of the capacitor C3, the high-speed field effect transistor (FET) is turned on and the NPN transistor (Q1) is in its off-state; no current flows through the $CuSO_4$ solution. Discharge is enabled when Q1 is turned on, and the FET off, but actually occurs only when contact is made in the manometric tube. At this time, current flows through the $CuSO_4$ solution and oxygen is released. Both transistors are controlled by a flip-flop (2 NOR gates). The voltage follower (VF) couples C3 with two voltage comparators (VC1 and VC2), which compare the actual voltage of the capacitor with two preset voltages (the minimum and maximum voltages V_{min} and V_{max}), both determined by the resistor bridge (R6, R7 and R8). The discharge of a 1-μF high-stability capacitor by 17.22 V releases 1 nl oxygen. The two voltage comparators control the state of the flip-

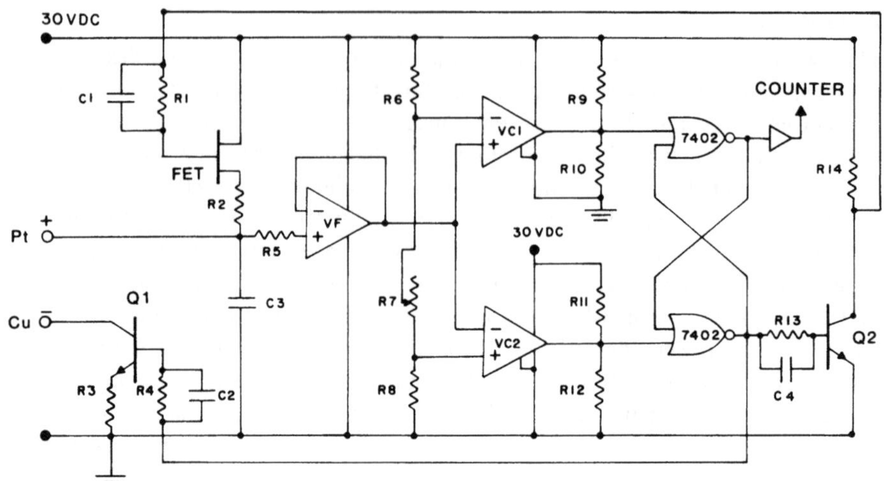

Fig. 7.4. Wiring diagram for the oxygen circuit. R1 = 10 MΩ; R2 = 333 Ω, 2 W; R3 = 1–10 kΩ adjustable; R4, R10, and R12 = 10 Ω; R5 = 10 kΩ; R6 = 2.8 kΩ; R7 = 10 kΩ adjustable; R8 = 3.9 kΩ; R9 and R11 = 5.6 kΩ; R13 = 100 Ω; R14 = 1.24 kΩ, 1 W; Q1 and Q2 are 2N222A; FET is 2N4392; VF is LM301; VC1 and VC2 are LM311.

flop. Besides enabling the discharge, Q1 also tends to maintain a constant discharge current. Changing the resistance of R3 permits adjustment of the current intensity to appropriate levels for producing pure oxygen (see Sect. V.2). This is an important adjustment when the microrespirometer is used at low temperatures.

The circuit then functions as follows: The charged capacitor that communicates with the cell of electrolysis discharges or remains charged depending on whether or not the microanode makes contact with the $CuSO_4$ solution. During discharge, the voltage of the capacitor rapidly reaches the minimum voltage. At this moment, comparator VC2 sets the flip-flop, which terminates the discharge by blocking Q1, and starts the charge by turning on the FET through which current flows until C3 is charged again to V_{max}. At this moment VC1 terminates the charge by resetting the flip-flop, which turns off the FET and turns on Q1. This completes the cycle of operations, a new one begins, and the cycling continues as long as the microanode remains in contact with the $CuSO_4$ solution. Each discharge produces an electrical pulse (5-V DC square wave), which is accumulated in a digital counter. A digital printer prints the content of the digital counter at preset time intervals.

4. Calibration

Calibration is based on the ideal gas law. The air space in the respiratory chamber is expanded by a known volume (constant number of moles) and the amount of electrolytic oxygen necessary to restore the initial pressure is then recorded.

During calibration (Fig. 7.5) the microrespirometer is enclosed in a large compensating chamber (an Erlenmeyer flask of 500-ml capacity) and communicates with an accurately calibrated capillary tube partially filled with mercury. By moving a stainless steel rod within the mercury reservoir the position of the thread of mercury inside the capillary may be changed. Displacement of the mercury from mark 1 to 2 expands the air space above the meniscus of the $CuSO_4$ solution by a volume equal to that enclosed in the capillary between the two marks. As a result, the pressure inside this air space drops and the $CuSO_4$ solution rises above its initial level of equilibrium, just below the microanode. Oxygen is then produced and the number of discharges (N) necessary to restore the initial pressure is recorded. The volume of oxygen produced is then equal to the imposed change in volume. The compensating chamber is supplied with three stopcocks: stopcock 1 communicates with the respiratory chamber; stopcocks 2 and 3 permit rapid flushing of the compensating chamber with controlled mixtures of gases. The calibration device is immersed in a temperature-controlled water bath (±0.01°C).

The capillary volume enclosed between the two marks has been determined by measuring the distance (D) between the two marks with an optical comparator (Gaertner Scientific Co.) and the mean cross-sectional area (A). The latter

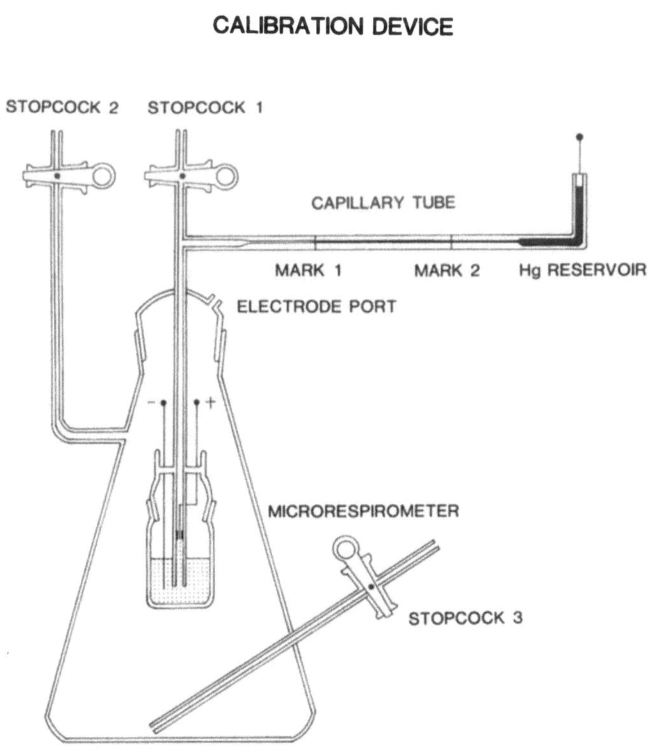

Fig. 7.5. Calibration apparatus.

Table 7.1. Experimentally Determined Coulometric Coefficients

	2.8°C	5°C	10°C	15°C
Coulometric coefficient	57614	57668	58403	57443
Standard deviation	±607	±607	±494	±754
Coefficient of variation (%)	1.05	1.05	0.85	1.31
Relative deviation from theory (%)	−0.79	−0.70	+0.57	−1.08

was determined by measuring the length and weight of threads of mercury within the marks.

At the beginning of calibration, the manometer and the thermobarometer are flushed with pure oxygen, so that errors due to dissolution and diffusion of oxygen are eliminated. Before the gas volume is expanded, care must be taken that the $CuSO_4$ meniscus is at its reference level, just below the platinum microanode, and that the calibration device is in thermal equilibrium with the surrounding water bath. The imposed volume changes are reduced to STPD. Measurements are made with a saturated $CuSO_4$ solution at 10 different randomized temperatures ranging from 2.8° to 40°C.

Instead of directly comparing the imposed change in volume with the volume of oxygen produced, we compute the coulometric coefficient k, which is the ratio of the imposed volume change to the total quantity of electricity involved, Eq. (7.2). The experimental k is then compared to the theoretical value of k. This procedure facilitates comparison of calibration data obtained at different temperatures. The mean experimental value of k relative to the theoretical value of k is an estimate of the effect of systematic errors (accuracy), whereas the reproducibility of k is an estimate of the overall effect of random errors in the calibrating process (precision). Since the calibration process itself might introduce additional errors not normally present in the automatic operations of the microrespirometer, the estimated reproducibility does not represent the precision of the coulometric microrespirometer.

The mean values of 10 determinations, their standard deviations, coefficients of variation, and relative deviations from the theoretical value of k for each temperature are shown in Table 7.1. The relative deviations from the theoretical value of k are on the order of 1%. The accuracy of the coulometric method is therefore of the same order of magnitude.

The long-term stability of calibration depends upon the constancy of the capacitance and the voltage change applied to the capacitor and upon the composition of the electrolytic solution. The performance of the electronic circuit used in this laboratory has not changed significantly over a period of several years. We checked whether the accumulation of H_2SO_4 has a significant effect on the calibration during the course of long-term experiments by repeating the calibrations with a saturated $CuSO_4$ solution containing 3% H_2SO_4: no signifi-

20°C	25°C	30°C	35°C	37.5°C	40°C
57718 ±571	57446 ±666	58516 ±571	58013 ±561	57941 ±492	57693 ±537
0.99	1.16	0.98	0.97	0.85	0.93
−0.61	−1.08	+0.76	−0.10	−0.23	−0.65

cant differences were found with respect to the pure $CuSO_4$ solution, although the oxygen uptake at the cathode increased.

The coefficients of variation reflect the reproducibility of the calibration process, rather than the precision of the method. In order to estimate the actual precision of the microrespirometer, we recorded the regular and constant oxygen uptake of copper wire (1.5 mg) placed in 0.3 ml saturated $CuSO_4$ solution containing 2% H_2SO_4. Such a preparation consumes on the average 310 nl/h. Since the standard deviation of 24 consecutive hourly values is 16 nl, the coefficient of variation is 5.16%. Assuming the rate of oxygen uptake is constant, this would mean that the volume of the respiratory chamber (4.5 ml) is maintained constant within ±16 nl or 3.5×10^{-4}%. These residual volume variations are equivalent to and may, in fact, be caused by nonsynchronous temperature fluctuations of ±0.001°C in the air spaces of the respiratory chamber and the thermobarometer.

The maximum rate of oxygen uptake measurable by this technique is limited by the maximum rate at which oxygen can be produced. The rate at which oxygen is produced is independent of the capacitance (Heusner 1970). Its maximum is determined by the resistance of the charge and discharge circuits. The resistance in the latter case is mainly determined by the resistance of the detecting microanode. Experimental determinations show that the maximum rate is about 1 ml/h.

IV. Chronocoulometric Respirometry

The chronocoulometric method is most suitable for recording oxygen consumption in insects that require larger respiratory chambers (honeybees, bumblebees, wasp, hornets, beetles, moths, etc.). In Fig. 7.6 the closed-system respirometer is shown; it consists of the respiratory chamber, the thermobarometer, the manometer, and the electrolytic cell. The glass respiratory chamber is conformable to the size, form, and natural habitat of the animal. The maximum relative error $\Delta V/V_{O_2}$ on the measured volume of oxygen is expressed in terms of the volume of the respiratory chamber V_R and the change in temperature $\Delta T°$

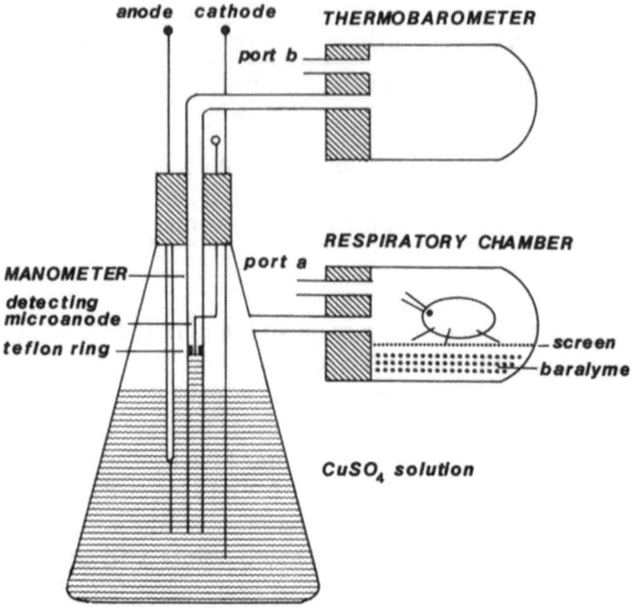

Fig. 7.6. Chronocoulometric respirometer. The entire apparatus is immersed in a temperature-controlled water bath. Long Tygon tubes extend ports a and b out of the water bath (not shown).

during the measurement; it therefore provides guidelines for the construction of a specific respiratory chamber (Heusner 1965):

$$\frac{\Delta V}{V_{O_2}} = \frac{V_R}{V_{O_2}} \alpha \Delta T^\circ \tag{7.9}$$

where $\alpha = 1/273$. For example, if $V_{O_2} = 1$ ml, $V_R = 50$ ml, and $\Delta T^\circ = 0.1^\circ C$, the maximum relative error would be 2%. In fact, the error is less because of the temperature compensation of the thermobarometer.

The respiratory chamber communicates with the air space of the electrolytic cell through a rubber stopper by means of Tygon tubing. Ports a and b are connected to long Tygon tubes that extend out of the water bath and are hermetically sealed with short glass rods and immersed in the water bath during the measurement process. Pressure equilibration of the immersed respiratory chamber is ensured when ports a and b are open to the atmosphere prior to the measurement of oxygen consumption. The respiratory chamber contains Baralyme for CO_2 absorption. A plastic screen separates the animal from the Baralyme.

The thermobarometer eliminates the effect of changes in barometric pressure and reduces the effect of residual temperature fluctuations in the water bath. Complete compensation would be obtained if the temperature variations were synchronous in the thermobarometer and the respiratory chamber and if thermal deformations of both chambers were the same.

The electrolytic cell is formed by a 250-ml Erlenmeyer flask containing a platinum anode and a copper cathode. The anode consists of a platinum wire, 20 mm long and 0.5 mm in diameter, sealed in a glass tube. The copper cathode is a Teflon-insulated copper wire, 2 mm in diameter, the last 40 mm of which are immersed into the $CuSO_4$ solution. It is important to keep the copper wire insulated at the air–solution interface to prevent its oxidation.

The manometer is a glass tube that projects into the $CuSO_4$ solution at one end and communicates with the thermobarometer at the other end. The manometer contains the detecting platinum microelectrode and the Teflon ring.

The respirometer is immersed in a constant-temperature water bath (±0.02°C) and functions as follows: As the pressure in the air space in the respiratory chamber decreases, the depth of the $CuSO_4$ meniscus increases and contact between the detecting microanode and the $CuSO_4$ solution is broken. This in turn causes a constant current to flow through the electrolytic solution during a preset time (1 sec). The oxygen pulse released at the anode pushes the meniscus back into contact with the detecting microanode. When this contact is again broken the animal has consumed an amount of oxygen equal to that released during the oxygen pulse. Note that the effect of circuit closing in this system is the reverse of that described previously for the capacitive coulometric system.

The wiring diagram of the pressure-detecting circuit and the oxygen pulse generator is presented in Fig. 7.7. When the contact between the detecting microanode and the $CuSO_4$ solution is open (see also Fig. 7.6), the CMOS driver (CD4050) sets the flip-flop (7400 NAND-gates). This enables (1) the clock pulses

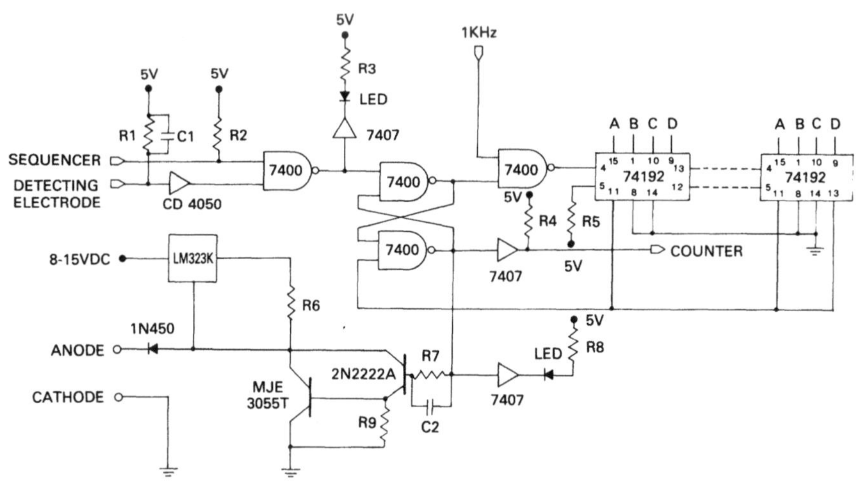

Fig. 7.7. Wiring diagram of the pressure-detecting circuit and the oxygen pulse generator. R1 = 10 MΩ; R2, R3, R4, R5, R7, and R8 = 300 Ω; R6 is adjustable resistor; R9 = 1000 Ω; C1 and C2 = 35 pF. Four down-counters 74192 are used.

(1 kHz) to pass the 7400 NAND-gate and decrement the preloaded four-digit down-counter (74192), and (2) a constant current to flow through the electrolytic cell by turning off the transistor MJE3055T, which is in parallel with the electrolytic cell. When the down-counter is emptied it resets the flip-flop, causing another pulse to be accumulated in a digital counter. A digital printer records the total number of pulses at preset time intervals. Current flow through the electrolytic cell is terminated when the transistor MJE3055T is enabled at the end of the pulse. Current ceases to flow if the oxygen released during one current pulse is sufficient to bring the meniscus back into contact with the microanode; if not, a new oxygen pulse is produced. The volume of oxygen V_{O_2} (in μl) released during one pulse is

$$V_{O_2} = (0.0581 \cdot N \cdot I)/f \qquad (7.10)$$

where I is the current intensity in milliamperes, N is the count of the preloaded down-counter, and f is the frequency of the clock pulse in hertz. The voltage regulator LM323K functions as a current regulator. The current intensity is determined by the resistor R6 or the output of the voltage regulator (Ohm's law).

1. Calibration

The respirometer is calibrated by filling the respiratory chamber with water and sampling known volumes from it with a syringe. The volume of the water sample is determined by weighing and is then compared to the volume of oxygen produced to restore the initial pressure. The replacement of air with water avoids the difficulty of accurately withdrawing small air samples from the respiratory chamber. Care must be taken to make sure the $CuSO_4$ solution is just below the detecting microanode at the beginning of the withdrawal. This is achieved by withdrawing the water just after a bout of electrolysis.

Calibrations have been performed at 5°, 15°, 25°, and 35°C with current intensities of 0.062, 0.165, and 0.350 A. A saturated $CuSO_4$ solution containing 0.1% $FeSO_4$ and 10% CuO was used. The results are expressed as the relative difference between the experimental and theoretical coulometric coefficients. The mean relative deviation of 10 determinations in each group is shown in Table 7.2. Unacceptably large deviations were found with a current of 0.062 A. This is due to the low anodic current density. Indeed, if the anodic surface is reduced by shortening the immersed portion of the platnium anode, relative errors of less than 1% are obtained.

There is a systematic trend, although negligible in practice, for the coulometric measures to be larger than the withdrawn volumes. This may be due either to a current yield of oxygen smaller than 100% or to oxidation of copper at the cathode. Comparison of the actual copper deposited at the cathode with the theoretical amount computed by Faraday's law also reveals a systematic

Table 7.2. Mean Relative Deviations (%) from the Theoretical Value of the Experimentally Determined Coulometric Coefficients[a]

Theoretical value					
I	acd	5°C	15°C	25°C	35°C
0.062	0.198	−9.00	−8.02	−4.43	−11.20
0.165	0.525	−0.29	−0.31	−1.62	−1.12
0.350	1.115	−0.37	+0.63	−0.44	+0.05

[a]Mean values of 10 determinations.

[b]*I*, current intensity in amperes; acd, anodic current density in amperes per square centimeter.

discrepancy in which the computed deposition is greater than the deposition actually measured. The small discrepancy (less than 2%) is due to the disappearance of copper either directly through dissolution in the $CuSO_4$ solution or indirectly through oxidation. These results seem to indicate that oxidation of copper is responsible for the consistently higher coulometric measure. Recording of oxygen uptake in the respiratory chamber without an animal confirms that a negligible volume of oxygen is consumed in the electrolytic cell. From these various tests we conclude that the accuracy of the chronocoulometric measure is better than 2%.

The sensitivity is given by the smallest detectable change in volume in the respiratory chamber. The gasometric sensitivity is mainly determined by the sharpness of the tip of the detecting microanode and its position relative to the Teflon ring. In a respiratory chamber of 350-ml volume, changes smaller than 1 μl can be detected, so that in practice the precision of the oxygen measurement is determined by the volume of oxygen produced during a pulse (i.e., the "unit volume"). Usually this unit volume is preset to represent about 1% of the minimum recorded volume of oxygen. This is accomplished by adjusting the current intensity (R6, Fig. 7.7) or the duration of the down-count pulse, *N*.

V. Sources of Error

The accuracy and precision of a coulometric measure depend on how effectively systematic errors, due to side reactions at the anode and cathode, and random errors in the gasometric detection can be reduced or eliminated. Calibration of both the capacitive microrespirometer and the chronocoulometric device has shown that these side reactions can be effectively controlled. Although the cathodic sources of errors are the same in both respirometers, their effects on the measure are opposite, because of the different locations of the cathodes. In the capacitive microrespirometer the cathode is located in the thermobarometer. In the chronocoulometric device it is located in the respiratory chamber. The effect of anodic side reactions is the same in both systems.

1. Cathodic Side Reactions

Practically all systematic errors originating at the cathode involve the release of hydrogen or oxygen uptake. Oxygen uptake at the cathode decreases the volume of the thermobarometer, and consequently the actual oxygen uptake in the respiratory chamber is underestimated. Hydrogen release is a source of error that is easily suppressed by adjusting the cathodic current density. In the chronocoulometric device, cathodic oxygen uptake will lead to an overestimate of the actual biological oxygen consumption. The following side reactions cause oxygen uptake: (1) dissolution of copper in the $CuSO_4$ solution with formation of Cu_2SO_4, which is then oxidized (Bockris and Enyo 1962); (2) copper deposition involving the formation of Cu_2SO_4, which in the presence of oxygen is oxidized to $CuSO_4$ (Richards et al. 1900); and (3) oxidation of the deposited copper.

In the microrespirometer, the effect of these side reactions becomes negligible when the volume of the thermobarometer is considerably larger than that of the respiratory chamber. If all the released copper is oxidized and reappears in newly regenerated $CuSO_4$, then the same amount of oxygen would be consumed at the cathode as is produced at the microanode, and the relative error cast on the actual oxygen uptake would be equal to the volume ratio between the respiratory chamber and the thermobarometer. In practice, since copper is observed to be deposited on the cathode, the relative error is actually smaller. The existence of these sources of error has been experimentally confirmed by directly measuring the oxygen uptake at the cathode and by recording a constant oxygen uptake of a relatively constant oxygen consumer, such as a small colony of yeast (*Rhodotorula minuta*), with different volumes of the thermobarometer. As predicted, the recorded oxygen uptake of the consumer becomes smaller when the volume of the thermobarometer decreases. When the volume ratio of V_R to V_T is 1:20 the maximum relative error of the recorded oxygen consumption is 5%. This source of error is negligible in the chronocoulometric device when the surface area of the copper cathode is reduced to a minimum to eliminate hydrogen production.

Sulfuric acid increases the oxygen uptake at the cathode. An increase in H_2SO_4 concentration during the experiment is prevented by adding CuO to the $CuSO_4$ solution. The CuO reacts with the H_2SO_4 to regenerate $CuSO_4$.

2. Anodic Side Reactions

Improper anodic current densities resulting in the production of ozone, hydrogen peroxide, persulfuric acid, etc., cause significant but constant deviations (up to 10%) from Faraday's law (see Potter 1956, Table 2). These undesirable side reactions divert current from the electrolytic production of oxygen, thereby decreasing the current efficiency. As a result, a larger quantity of electricity is used, and the actual oxygen uptake in the respiratory chamber is overestimated. No significant deviations from Faraday's law were found with current densities from 0.5 to 1.2 A/cm^2. In the microrespirometer current density is adjusted by the resistance of the discharging circuit (R3, Fig. 7.4). In the chronocoulometric device current density is adjusted by the size of the platinum anode.

Ozone is a normal side product of the electrolytic oxygen production (Woodland 1973). The trace amounts of ozone can be suppressed without altering the calibration by adding 1 g/liter of $FeSO_4$ to the $CuSO_4$ solution (Heusner and Petrovic 1964).

Deviations from Faraday's law may also be observed in the presence of electrical leaks when not all of the measured current flows through the $CuSO_4$ solution. Dissolution of nascent oxygen in the $CuSO_4$ solution, with eventual diffusion out of the respiratory chamber into the compensating vessel, can cast considerable errors on the measure. This error becomes negligible if the partial pressure of oxygen in the $CuSO_4$ solution is in equilibrium with that in the respiratory chamber and in the thermobarometer.

VI. Control Circuits

The peripheral electronic circuits that are used to automatically print out oxygen consumption measurements are shown in Fig. 7-8. The timer circuit consists of three down-counters (74192) that are preloaded with a number corresponding to the desired duration of a measurement in minutes. The down-counters are emptied by 1-min pulses that are produced by a quartz clock (not shown). When the counters are empty, the sequencer circuit is activated, which controls latching, printing, and resetting of the oxygen counters.

The sequencer circuit allows sequential, on-line printing and resetting of oxygen consumption measurements from two respirometers. A +5-V DC pulse from the timer circuit sets the flip-flop (two NOR gates, 7400), which activates a sequence of the following events: latching, printing, and resetting of the DM8552 counter of circuit 1 followed by the same sequence for circuit 2.

The counter circuit consists of a series of tristate decade counters (DM8552) that count the number of unit volumes of oxygen produced by each oxygen circuit. The wiring diagram is shown for only one channel (top), which allows counting from 1 to 999. More counters may be added to increase the range. The counters for the second channel are schematically illustrated at the bottom of the figure, but are wired according to the connections shown at the top. The 4-bit output of each counter is bussed to a Datel Model DPP-Q7 printer. Notice that the same bus lines are used by both channels. This is made possible by the sequential control of latching, printing, and resetting for each channel by the sequencer circuit.

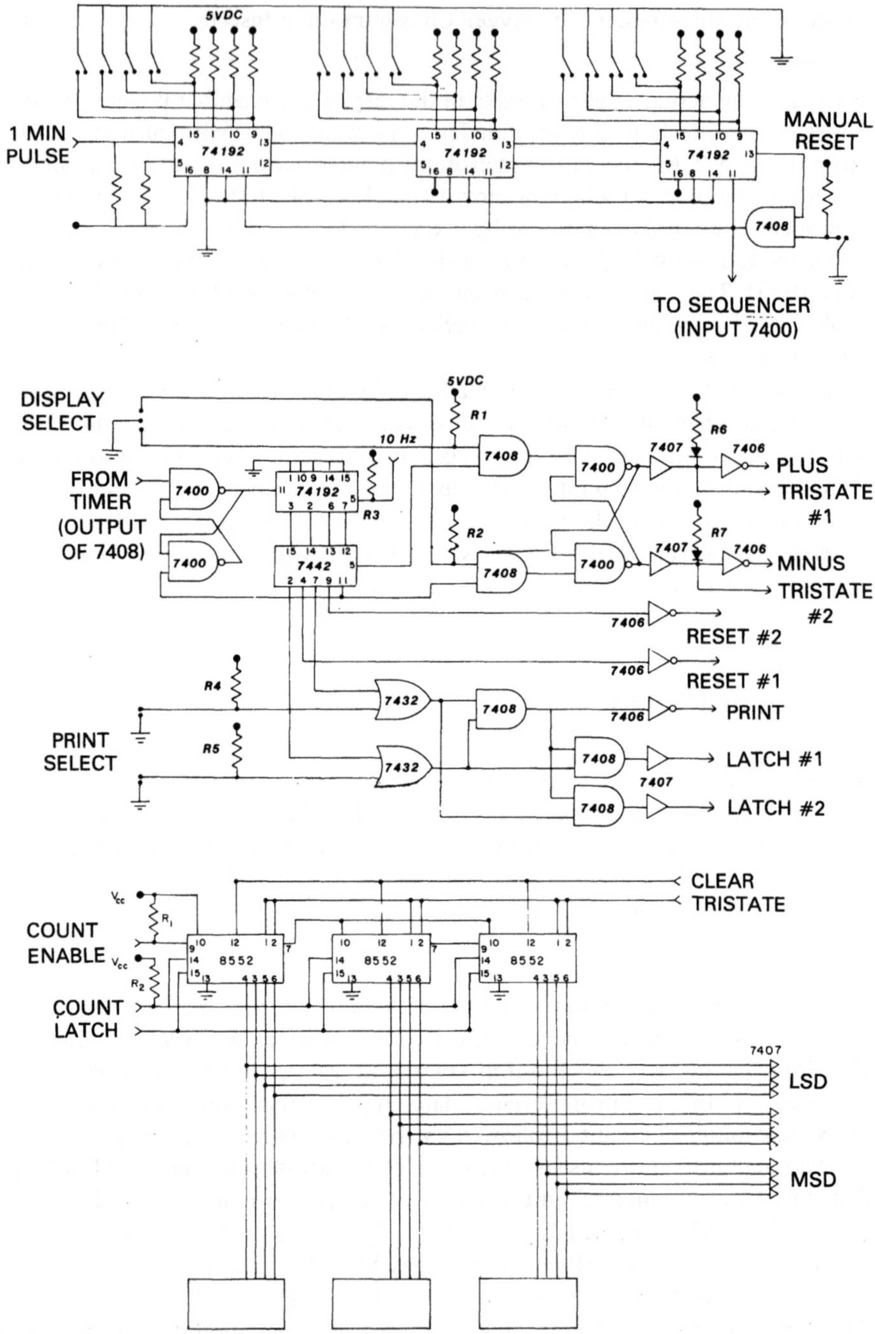

Fig. 7.8. Wiring diagram of the timer, sequencer, and counter circuits. Timer circuit: all resistors are 1 kΩ. The 1-min pulses are provided by a quartz clock. Sequencer circuit: R1, R2, R3, R4, and R5 = 1 kΩ; R6 and R7 = 220 Ω. Counter circuit: the circuit is shown in detail for one counter only; R1 and R2 = 1 kΩ. More counters can be connected to the bus line. LSD, least significant digit; MSD, most significant digit.

VII. Examples of Application of Coulometry

A 7-day recording of oxygen consumption in an axenic female fruit fly at 25°C is shown in Fig. 7.9. Oxygen consumption was recorded by capacitive coulometry. The respiratory chamber containing food (5% glucose and 5% agar-agar) and Baralyme was autoclaved before the experiment. Control measurements showed that the medium did not consume oxygen. During the recording in Fig. 7.9 the fly was subjected to a light-dark cycle (12 h each) by means of a fiber-optic system, which was shown not to interfere with the measuring process during control tests. The fly displayed a bimodal daily rhythm in oxygen consumption, the two peaks maintaining a constant phase relationship with the transitions of the light-dark cycle. However, since oxygen consumption increased consistently before the change from dark to light and vice versa, the

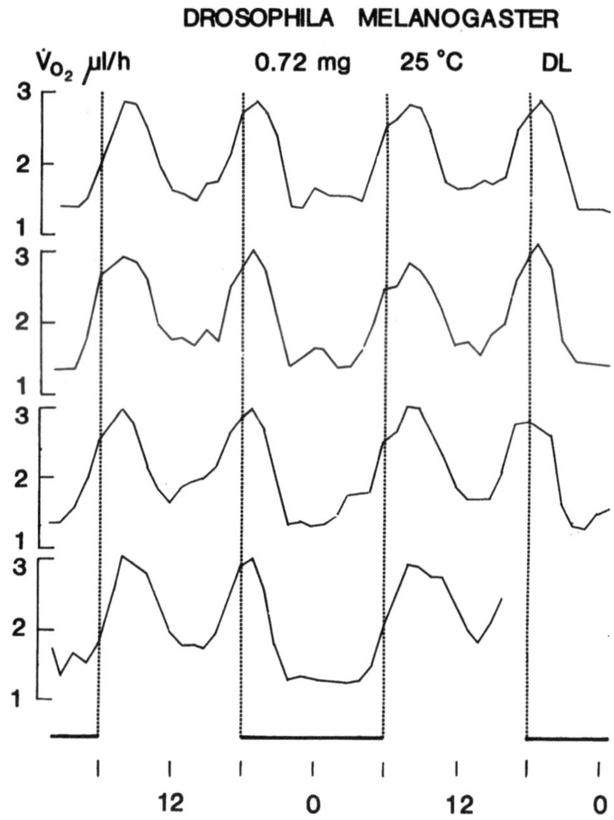

Fig. 7.9. Seven-day recording of oxygen consumption in an axenic female fruit fly (*Drosophila melanogaster,* 0.72 mg, 25°C) subjected to a 12:12 dark-light cycle. The scotophase is indicated by the continuous line.

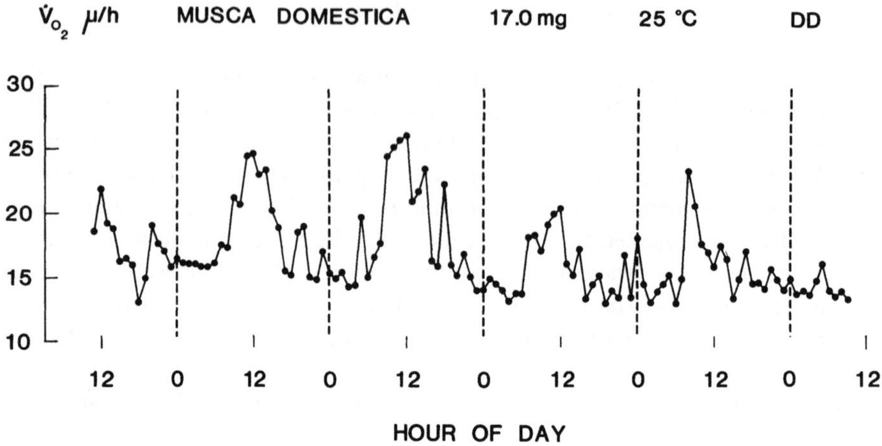

Fig. 7.10. Five-day recording of oxygen consumption in a house fly (*Musca domestica,* 17.0 mg, 25°C) placed in continuous darkness.

increase in oxygen consumption is not an immediate metabolic response to the light–dark transitions.

A 5-day recording of oxygen consumption by capacitive coulometry in a larger insect, the house fly, is presented in Fig. 7.10. During this experiment the fly was not fed and was kept under constant darkness. This recording reveals a daily monophasic cycle, the high phase occurring during daytime.

A 3-day recording of oxygen consumption in a cockroach is shown in Fig. 7.11. It was also deprived of food, but was subjected to the natural light–dark cycle in the laboratory. The oxygen consumption was recorded by chronocoulometry. This animal also displayed a very definite daily metabolic cycle, but in this case the high phases occur during the night.

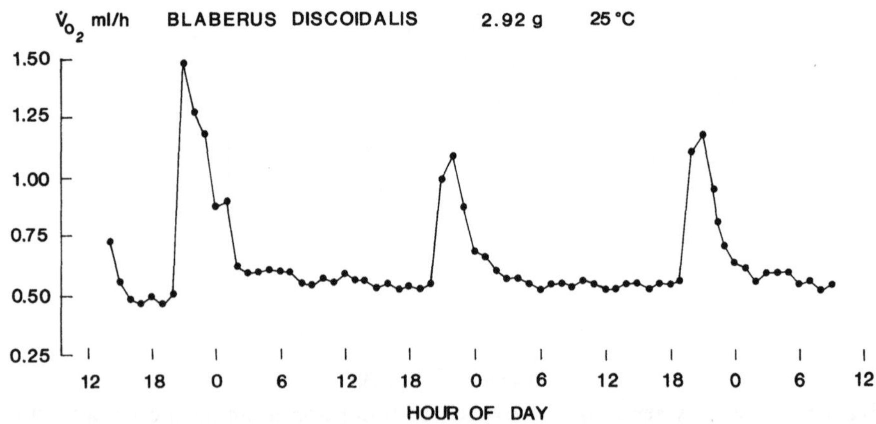

Fig. 7.11. Three-day recording of oxygen consumption in a cockroach (*Blaberus discoidalis,* 2.92 g, 25°C) subjected to the natural change of light and dark.

VIII. Concluding Remarks

The simplicity and sensitivity of the coulometric respirometer, combined with the accuracy of measurement over long periods of time, make the coulometric principle ideal for determining long-term as well as acute effects of chemicals, nutrients, gas composition, genetic manipulations, etc. on the energy metabolism of insects at all stages of development. The digital technique may be improved by directly interfacing the counting circuits with a microprocessor, thereby completely automating the measuring process, data acquisition, and reduction.

References

Adam NK (1964) The chemical structure of solid surfaces as deduced from contact angles. In contact angle, wettability, and adhesion. Adv Chem Ser **43**: 52–56

Bennet-Clark TA (1932) A method for automatically recording the oxygen intake of living tissues. Sci Proc Ro Dublin Soc **20**:281–291

Bockris JO'M, Enyo M (1962) Mechanism of electrodeposition and dissolution processes of copper in aqueous solutions. Trans Faraday Soc **58**:1187–1202

Brown Jr FA (1954) Simple automatic continuous recording respirometer. Rev Sci Instrum **25**:415–417

Capraro V (1953) A new method of measuring oxygen consumed in the metabolism of small animals. Nature **172**:815

Chase AM, Unwin DM, Brown RHJ (1968) A simple electrolytic respirometer for the continuous recording of oxygen consumption under constant and natural conditions. J Exp Biol **48**:207–215

DeBoer SP (1929) Respiration of Phytomycetes. Rec Trav Bot Neerl **25**:117–329

Fernandes DS (1923) Aerobe und anaerobe Atmung bei Keimlingen von *Pisum sativum*. Rec Trav Bot Neerl **20**:103–256

Fourche J (1964) Un respiromètre électrolytique pour l'étude des pupes isolées de drosophiles. Bull Biol Fr Belg **98**:475–489

Hayward JS, Nordan HC, Wood AJ (1963) A simple electrolytic respirometer for small animals. Can J Zool **41**:63–68

Heusner AA (1965) Sources of error in the study of diurnal rhythm in energy metabolism. In: Aschoff J (ed) Circadian clocks, Proceedings of Feldafing Summer School, Sept 1964. North-Holland, Amsterdam

Heusner AA (1970) Long term numerical recording of very small oxygen consumptions under sterile conditions. Respir Physiol **10**:132–150

Heusner AA, Jameson Jr. EW (1981) Seasonal changes in oxygen consumption and body composition of *Sceloporus occidentalis*. J Comp Biochem Physiol **69A**:363–372

Heusner AA, Lavoipierre MMJ (1973) Effet énergétique du repas sanguin chez *Aedes aegypti*. C R Acad Sci D (Paris) **276**:1725–1728

Heusner A, Petrovic A (1964) Appareil de culture d'organes en milieu liquide continuellement oxygéné. Med Electron Biol Eng **2**:381–385

Heusner A, Ruhland ML (1959) Dispositif permettant l'enregistrement simultané de la consommation d'oxygène et de l'activité chez des poissons de petite taille. J Physiol (Paris) **51**:580

Heusner A, Stussi TH (1964) Métabolisme énergétique de l'abeille isolée: Son rôle dans la thermorégulation de la ruche. Insect Soc **11**:239–266

Heusner A, Stussi TH, Dreyfus E (1965) Application de la coulométrie à la mesure de la consommation d'oxygène. Med Electron Biol Eng **3**:39–56

Heusner AA, Lavoipierre MMJ, Bond DC (1973) Etude cinétique de l'effet métabolique d'un repas sanguin chez *Aedes aegypti*. C R Acad Sci D (Paris) **277**:2017–2020

Heusner AA, Hurley JP, Arbogast R (1982) Coulometric microrespirometry. Am J Physiol **343**:R185–R192

Kayser CH, Heusner AA (1967) Le rythme nyctéméral de la dépense d'énergie. J Physiol (Paris) **59**:3–116

Klekowski RZ, Zajdel JW (1972) Capacity electrolytic respirometer KZ-CER-01T with review and discussion of electrolytic respirometry. Pol Arch Hydrobiol **19**:475–504

MacFayden A (1961) A new system for continuous respirometry of small air breathing invertebrates under near normal conditions. J Exp Biol **38**:323–341

Moyat P (1957) Uber den Einfluss von Licht und Aktivitat auf endogene Stoffwechselrhythmen bei Kleinsaugern und Vogeln. Z Vergl Physiol **40**:397–414

Potter EC (1956) Electrochemistry principles and applications. Cleaver-Hume Press, London

Richards WR, Collins E, Heimrod GW (1900) Das elektro-chemische Aequivalent des Kupfers und des Silbers. Z Phys Chem **32**:321–347

Stussi TH (1967) Thermogenèse de l'abeille et ses rapports avec le niveau thermique de la ruche. Thèse Etat Science, Lyon

Stussi TH, Heusner A (1963) Variation nycthémérale de la consommation d'oxygène chez quelques espèces d'insectes. C R Soc Biol (Pairs) **157**:1509–1512

Swaby RJ, Passey BI (1953) A simple macrorespirometer for studies in soil microbiology. Aust J Agric Res **4**:334–339

Turner BD, Stevenson RA (1974) An electrolytic, digital recording, multichannel microrespirometer. J. Exp Biol **61**:321–329

Ulrich A (1940) Measurement of the respiratory quotient of plant tissues in a gaseous environment. Plant Physiol **15**:527–536

Wager HG, Porter FAE (1961) An apparatus for automatic measurement of oxygen uptake by electrolytic replacement of oxygen consumed. Biochem J **81**:614–618

Werthessen NT (1937) An apparatus for the measurement of the metabolic rate of small animals. J Biol Chem **119**:233–239

Winteringham FPW (1959) An electrolytic respirometer for insects. Lab Pract **8**:372–375

Wolbarsht ML, MacNichol Jr EF, Wagner HG (1960) Glass insulated platinum microelectrode. Science **132**:1309–1310

Woodland DJ (1973) The ozone problem in electrolytic respirometry and its solution. J Appl Ecol **10**:661–662

Woodland DJ, Clack DJ (1975) Simplified constant temperature electrolytic respirometer. Lab Pract **24**:518–521

Chapter 8

Techniques for Studying Na^+,K^+-ATPase

J. H. Anstee and K. Bowler

I. Introduction

Cells maintain, by active transport, a high intracellular K^+ and low intracellular Na^+ relative to the extracellular compartment. Hodgkin and Keynes (1955) showed that active transport of Na^+ and K^+ required cellular metabolism, and subsequently that it was dependent on the supply of "high-energy" phosphorylated compounds (Caldwell et al. 1960). Skou (1957) was the first to demonstrate that active transport might have an enzymatic basis. From crab nerve he isolated membrane fragments that possessed a Mg^{2+}-dependent adenosine triphosphatase (ATPase) enzyme activity that was stimulated by the simultaneous presence of Na^+ and K^+. It was significant that Skou found that the monovalent cation-stimulated activity was abolished by ouabain, an inhibitor of active Na^+ transport (Schatzmann 1953). The enzyme was described as the Na^+,K^+-dependent ATPase (E.C.3.6.1.3.).

It is not the purpose of this chapter to review the recent literature on this enzyme; this has been done by Glynn and Karlish (1975), Robinson and Flashner (1979), and Skou (1975). However, in order that the principles underlying the various methods used in studying Na^+,K^+-ATPase may be appreciated, it is necessary to describe its salient properties. *In vitro* Na^+,K^+-ATPase displays the following characteristics:

It is present in one or more membrane fractions of a tissue homogenate.
It preferentially hydrolyses ATP.
It has a requirement for Mg^{2+}.
It has a requirement for Na^+ plus K^+.

It is inhibited by cardiac glycosides.
Delipidation of membrane causes inactivation.

The enzyme is vectoral, spanning the membrane and in consequence presenting different sites at the inner (i) and outer (o) faces of the membrane. The catalytic center is situated at the i-face, and thus Mg^{2+} and ATP are required intracellularly. Two classes of ionic sites exist, one at the inner face (i-site) and one at the outer face (o-site). The requirement for Na^+ at the i-site is absolute, but the o-site will accept cations other than K^+ with the effectiveness, $K^+ > Rb^+ > NH_4^+ > Cs^+ > Li^+$ (Skou 1965).

In vitro membrane preparations of the enzyme exhibit activation curves for Na^+ (in the presence of K^+) and for K^+ (in the presence of Na^+) that are sigmoidal. This suggests that more than one Na^+ is bound at the i-site and more than one K^+ is bound at the o-site during ion activation (Garay and Garrahan 1973). The ionic conditions that give half-maximal and maximal activation of the enzyme are cited by Bonting (1970) for preparations from various sources, and show close correspondence. The enzyme is half-maximally activated by 10–12 m*M* Na^+ (at 5–10 m*M* K^+), and 60–100 m*M* Na^+ results in maximal activation. Higher concentrations of Na^+ are usually inhibitory, presumably because of competition with K^+ for the o-site. Half-maximal activation of the enzyme requires 1–1.5 m*M* K^+ (at 60 m*M* Na^+), and 5–10 m*M* K^+ results in maximal activation. Above 40 m*M* K^+, the enzyme is inhibited. The multiple binding of cations at the i- and o-sites is supported by the report that, in the red blood cell enzyme, 3 Na^+ are exchanged for 2 K^+ for each ATP molecule hydrolyzed (Glynn 1968).

In the absence of Na^+ no ATPase activity occurs, but a K^+-dependent ouabain-inhibitable phosphatase can be demonstrated, when artificial substrates (e.g., *p*-nitrophenyl phosphate) can be hydrolyzed. Under these conditions no ions are transported (Garrahan and Rega 1972). With Na^+ but no K^+ present the enzyme can be readily phosphorylated by ATP (Post et al. 1965) and dephosphorylation is very slow.

The enzyme requires Mg^{2+} in addition to ATP for catalytic activity. Maximal activity is usually found when the $[Mg^{2+}]:[ATP]$ ratio is 1; consequently, it is thought that a MgATP complex forms the substrate. The currently held view of the reaction sequence is developed from a model proposed by Albers (1967) (E_1 and E_2 are different states of the enzyme, P is phosphate, and P_i is inorganic phosphate):

1. Formation of the phosphorylated intermediate:

$$E_1 + ATP \underset{Na^+}{\overset{Mg^{2+}}{\rightleftharpoons}} E_1 \sim P + ADP$$

2. Probable step at which high-affinity Na^+ i-sites become outward oriented and change to K^+ affinity sites; Na^+ unloads, K^+ loads:

$$E_1 \sim P \rightleftharpoons E_2 - P$$

3. K^+ stimulation of discharge of phosphorylated intermediate:

$$E_2 - P + H_2O \overset{K^+}{\rightleftharpoons} E_2 + P_i$$

4. Reverse of the cationic i- and o-sites; restoration of initial orientation and affinities:

$$E_2 \rightleftharpoons E_1$$

The ouabain-binding site is at the o-face of the membrane. Binding of cardiac glycosides is modulated by ligands, and its rate closely parallels the rate of inhibition of the microsomal Na^+,K^+-ATPase (Albers et al. 1968; Allen et al. 1970). The results obtained suggest that ouabain binds to the $E_2 - P$ intermediate of the enzyme. This forms in the presence of Mg^{2+}, ATP, and Na^+, and these conditions favor binding and ouabain inhibition (Albers et al. 1968; Sen et al. 1969; Skou et al. 1971); but see also Sect. III.4B. Adding K^+, which discharges the $E_2 - P$ intermediate, reduces both binding and inhibition by ouabain (Matsui and Schwartz 1966). It is important to stress that K^+ reduces the rate of ouabain binding and of enzyme inhibition, even under what are otherwise optimal conditions (Lindenmayer and Schwartz 1973). This dependence of glycoside inhibition on ionic conditions is likely to be a consequence of the enzyme undergoing ion-dependent changes in conformation that are in some way related to ion translocation. This implies that the ouabain receptor is presented in certain conformational states only (Schwartz et al. 1975).

The Na^+,K^+-ATPase is always accompanied by other ATPases, and this complicates assay procedures. The commonest contaminant is Mg^{2+}-ATPase, an enzyme that requires Mg^{2+} only for activation. Its function is unknown but it is unlikely to be "uncoupled" Na^+,K^+-ATPase (Bonting 1970). In some preparations it is the major component of the ATPases present (e.g., liver plasma membrane), and it is usually insensitive to monovalent cations and ouabain. However, many workers have shown the enzyme in their preparations to be stimulated by Na^+ (Gilbert and Wyllie 1975; Proverbio et al. 1975; Tirri et al. 1979)—an activation insensitive to ouabain. Izutsu et al. (1974) also report a Mg^{2+}-ATPase that responds to the ionic strength of the assay. These effects on various Mg^{2+}-ATPases are realized under ionic conditions used in the demonstration of the Na^+,K^+-ATPase. Therefore, problems of interpretation can result if the appropriate isolation and assay procedures have not been followed, particularly when dealing with a Na^+,K^+-ATPase derived from a source not previously characterized for this enzyme.

The first demonstration of Na^+,K^+-ATPase activity in preparations from an insect was performed by Grasso (1967) on a nerve microsomal fraction from *Periplaneta americana.* The early reports of failure to demonstrate the enzyme in preparations from epithelial tissues were clearly the result of the application of inadequate methods. The first successful demonstration of the enzyme in membrane fractions from Malpighian tubules and hindgut was that by Peacock et al. (1972). Since then about 30 other reports have been published on about 20 insect species showing the presence of the enzyme in epithelial tissues. A summary of the published work to date on the Na^+,K^+-ATPases from insects and a tick is presented in Table 8.1.

Table 8.1. Properties of Na^+,K^+-ATPase from Various Insect and one Tick Species

Species	Tissue	Preparation	Specific activity	Assay temp °C
Locusta migratoria	Recta	Lyophilate	120	30
Locusta migratoria	Malpighian tubules	Microsomes	229	30
Locusta migratoria	Malpighian tubules	Microsomes	292	30
Locusta migratoria	Malpighian tubules	Microsomes	213	30
Locusta migratoria	Recta	Microsomes	105	30
Locusta migratoria	Recta	Microsomes	80	30
Locusta migratoria	Recta	Lyophilate	140	30
Periplaneta americana	Nerve cord	Microsomes	†	37
Periplaneta americana	"Brain"	Crude supernatant	2	24
Periplaneta americana	Nerve cord	Mitochondrial pellet	438	27
Periplaneta americana	Rectum	Lyophilate	500	–
Periplaneta americana	Antenna	Microsomes	250	30
Manduca sexta	Nerve cord	Microsomes	73	–
Manduca sexta	Head	Crude homogenate	‡	–
Hyalophora cecropia	Head	Crude homogenate	‡	–
Danaus plexippus	Head	Crude homogenate	‡	–
Calliphora erythrocephala	Eye	Crude homogenate	320	–
Calliphora erythrocephala	Brain	Crude homogenate	330	–
Musca domestica	Brain	Microsomes	1000	25
Aedes aegypti	Larvae	Mitochondrial pellet	80	37
Drosophila melanogaster	Imaginal discs	Homogenate	23	–

In the following sections the methods are presented that have been successfully used in the isolation, assay, and localization of Na^+,K^+-ATPase by workers using insects. Where they are understood, the principles underlying the methodology are discussed.

	Na^+ (m*M*)		K^+ (m*M*)			ATP (m*M*)	
pI_{50} ouabain	app. K_m	[max]	app. K_m	[max]	pH	app. K_m	Reference
–	–	–	–	–	–	–	Peacock (1979)
6.1	–	–	–	–	–	–	Anstee and and Bell (1975)
–	–	–	–	–	–	–	Donkin and Anstee (1980)
–	28	100	1	20	7.5	0.18	Anstee and and Bell (1978)
–	–	–	–	–	–	–	Peacock (1976a)
–	–	–	–	–	–	–	Peacock (1978)
6.0	–	–	–	–	–	–	Peacock (1981a)
6.4	–	100	5	20	7.6	–	Grasso (1967)
–	–	–	–	–	7.4	0.07	Piccione and Baust (1977)
–	–	–	–	–	–	–	Cheng and Cutcomp (1975)
5.3	19.6	120	2.7	10	7.05	–	Tolman and Steele (1976)
6.2	6.3	90	3.7	20	9.0	0.23	Norris and Cary (1982)
6.0	–	54	3.7	54	7.5	–	Rubin et al. (1980)
5.0	–	–	2.2	–	–	–	Vaughan and Jungreis (1977)
5.3	–	–	–	–	–	–	Vaughan and Jungreis (1977)
<3.0	6	–	2.2	–	–	–	Vaughan and Jungreis (1977)
5.75	30	100	2	15	7.6	–	Rivera (1975)
–	–	–	–	–	–	–	Rivera (1975)
5.6	20	–	1	–	–	0.5	Jenner and Donnellan (1976)
–	–	–	–	–	7.2	–	Yap and Cutcomp (1970)
–	–	–	–	–	–	–	Fristrom and Kelly (1976)

Table 8.1. (continued)

Species	Tissue	Preparation	Specific activity	Assay temp °C
Paragnetina media	Gill	Microsomes	630	37
Paragnetina media	Malpighian tubules	Microsomes	1087	37
Paragnetina media	Rectum	Microsomes	700	37
Glossina morsitans	Midgut	Lyophilate	120	30
Glossina morsitans	Midgut	Lyophilate	157	30
Glossina morsitans	Hindgut	Lyophilate	97	30
Sarcophaga nodosa	Midgut	Lyophilate	13	30
Sarcophaga nodosa	Recta	Lyophilate	79	30
Bombyx mori	Ileum	Lyophilate	0	30
Bombyx mori	Rectum	Lyophilate	0	30
Homorocoryphus nitidulus	Malpighian tubules and rectum	Microsomes	240	30
Blaberus craniifer	Rectum	Microsomes	86	30
Schistocerca gregaria	Rectum	Microsomes	75	30
Schistocerca gregaria	Rectum	Microsomes	67	30
Jamaicana flava	Rectum	Microsomes	18	30
Rhodnius prolixus	Follicle cells	Microsomes	40	37
Rhodnius prolixus	Follicle cells	Microsomes	90	37
Chironomus thummi	Salivary gland	Crude homogenate	†	37
Amblyomma hebraeum (tick)	Salivary gland	Microsomes	516	37

Specific activity in nanomoles Pi liberated per milligram protein per minute.
‡Quoted activity not convertible to specific activity.
†Quoted activity not convertible to an equivalent specific activity.
app. K_m = the apparent Michaelis constant (K_m) of the enzyme for the ligand.
[max] = concentration of ligand giving maximal enzyme activity.

II. Preparation of Tissue Fractions Containing Na^+,K^+-ATPase Activity

1. General Remarks

To demonstrate biochemically the presence of Na^+,K^+-ATPase activity in a tissue, it is first necessary to disrupt the cells, for only then can exogenous ATP gain access to the catalytic center of the enzyme. Intact cells are penetrated by ATP very slowly. Studies on the isolation of the enzyme from many vertebrate

	Na^+ (m*M*)		K^+ (m*M*)			ATP (m*M*)	
pI_{50} ouabain	app. K_m	[max]	app. K_m	[max]	pH	app. K_m	Reference
–	–	–	–	–	7.4	–	Kapoor (1980)
–	–	–	–	–	–	–	Kapoor (1980)
–	–	–	–	–	–	–	Kapoor (1980)
6.0	–	–	–	–	–	–	Peacock (1982)
–	–	–	–	–	–	–	Peacock (1981b)
–	–	–	–	–	–	–	Peacock (1981b)
–	–	–	–	–	–	–	Peacock (1981b)
–	–	–	–	–	–	–	Peacock (1981b)
–	–	–	–	–	–	–	Peacock (1981b)
–	–	–	–	–	–	–	Peacock (1981b)
6.5	–	–	3	20–25	7.4	0.14	Peacock et al. (1976)
–	–	–	–	–	–	–	Peacock (1977)
–	–	–	–	–	–	–	Peacock (1977)
–	–	–	–	–	–	–	Peacock et al. (1972)
–	–	–	–	–	–	–	Peacock et al. (1972)
–	–	–	–	–	–	–	Abu-Hakima and Davey (1979)
6.0	37	100	2.2	10	7.4	0.79	Ilenchuk and Davey (1982)
6.92	3.5	10	1.5	5	7.6	0.5	Schin and Kroeger (1980)
6.52	21	–	1.2	–	7.0	–	Rutti et al. (1980)

tissues have shown that particular care is necessary concerning the conditions used if the preparation is to be successful. The usual method is to disrupt the tissue by homogenization in a buffered isotonic medium containing ethylenediaminetetraacetate (EDTA), then isolate an enzyme-rich membrane fraction by differential centrifugation (Jørgensen 1974). This procedure involves two major problems. First, membrane fragments produced by homogenization form vesicles of various sizes; thus membranes possessing Na^+,K^+-ATPase activity may be distributed through several fractions and the yield may be low.

Second, it is usual for a proportion of the total Na^+,K^+-ATPase activity to be latent (Jørgensen and Skou 1971). This occurs because some vesicles are closed to the penetration of substrate and ligands. Such latent activity can be revealed by disruption of the vesicles with detergents. The methods used, and the degree of purification attempted, depend to a large extent on the requirement of the study. "Crude" tissue homogenates, without further purification, have a limited use in giving an approximate comparative level of enzyme activity in that tissue. They are of little value in studies that attempt characterization of the enzyme, where preparations of the highest purity possible should be used.

A major problem in using insect material is that only a relatively small mass of tissue is available from each individual; tissue from several or many individuals must be pooled for use. A further problem is that it is often difficult to obtain a tissue free from contamination with other tissues (Jenner and Donnellan 1976; Piccione and Baust 1977).

The commonest method used for tissue dispersion with insect material has been homogenization in a buffered medium with a tonicity raised by sucrose (Cheng and Cutcomp 1975; Fristrom and Kelly 1976; Grasso 1967; Kapoor 1980; Piccione and Baust 1977; Rivera 1975; Yap and Cutcomp 1970) or mannitol (Anstee and Bell 1975; Donkin and Anstee 1980; Peacock et al. 1972, 1976). Homogenization in distilled water or low ionic strength buffer has also been used (Abu-Hakima and Davey 1979; Peacock 1979; Rutti et al. 1980; Schin and Kroeger 1980; Tolman and Steele 1976). Sonication alone, or with homogenization, has also been used (Kapoor 1980; Rubin et al. 1980; Schin and Kroeger 1980).

The main procedural differences between various laboratories working with Na^+,K^+-ATPase from insect sources are in the methods used to disperse the tissue and obtain fractions for assay. It is not possible to appraise the various methods here, but all derive from techniques successfully developed for use with vertebrate tissues. Whatever the method chosen, ice-cold media should be used and the preparation should be kept at 0°–4°C during the isolation procedure. This protects the enzyme from denaturation and degradation by hydrolytic enzymes released. EDTA is also usually included in isolation media to chelate any trace metal ion that might be inhibitory.

2. Microsomal Na^+,K^+-ATPase from Malpighian Tubules of *Locusta*

A. Conventional Procedure

Details of a method for obtaining a microsomal preparation from *Locusta* Malpighian tubules are given to illustrate the general features of the isolation procedure. The method is based on that developed by Nakao et al. (1965) for mammalian brain.

Twenty adult locusts are used. They are taken in batches of four, and the entire gut is dissected out into ice-cold homogenization medium of the following composition: 250 m*M* mannitol, 5 m*M* EDTA, 0.1% deoxycholate in 30 m*M* histidine buffer pH 7.2.

The Malpighian tubules are rapidly dissected free of the mid- and hindguts and are then placed in 10 ml fresh, ice-cold homogenization medium in a glass homogenizer tube standing in an ice bath. Fresh medium is required for dissection after each batch of four locusts to minimize contamination by gut contents. The collected tubules are homogenized in a Potter-Elvehjem homogenizer with a glass tube and Teflon pestle (clearance 0.1–0.15 mm) at 1000 rpm. It is best to avoid the introduction of air into the medium by suction during the 10–15 passes of the plunger through the medium. The homogenizer should be surrounded by ice during homogenization.

The homogenate is then extracted with an equal volume of cold medium containing 5 m*M* $MgCl_2$, 10 m*M* EDTA, 4 *M* NaI (final NaI concentration is 2 *M*),

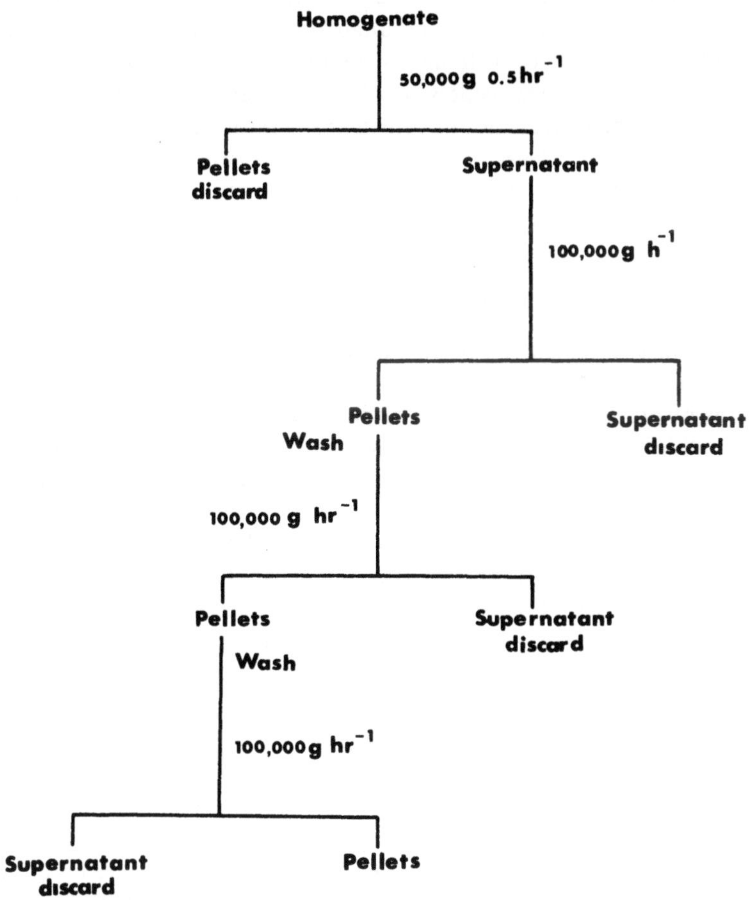

Fig. 8.1. Centrifugation steps involved in isolation of microsomal pellets following homogenization and extraction with NaI (see text). Washing medium contains 5 m*M* NaCl and 5 m*M* EDTA, pH 7.2.

(pH 7.2), and left to stand in ice for 30 min. The mixture is then diluted with cold deionized water so that the NaI concentration is reduced to 0.8 *M*.

The purpose of exposing membranes to deoxycholate and NaI is to unmask latent Na^+,K^+-ATPase and to solubilize other membrane-associated proteins. The Mg^{2+}-ATPase activity is particularly reduced by this procedure (Peacock et al. 1972).

The diluted homogenate is then centrifuged in an MSE Prepspin (rotor No. 43114-125) according to the schedule in Fig. 8.1. The final pellets are resuspended by hand homogenization in neutralized, ice-cold deionized water. For immediate use the final protein concentration is between 50 and 300 μg/ml, so the pellets are dispensed in about 20-30 ml. Preparations for storage should be made up to 2-5 mg/ml. Activity remains over several days if stored at 0°C; but repeated freezing and thawing at −20°C causes a progressive loss in activity.

Centrifugation procedures used in laboratories where insect tissues are used have required angle rotors. These are suitable where firm pellets result. They are also useful if large volumes of homogenate are in use, although in studies on insects this is unlikely. If loose pellets are obtained, then swing-out rotors are more suitable because the pellet is not disturbed as the rotor slows and stops. If a purification step requires a density gradient spin, then swing-out rotors are essential. Thus far no reports are available on the use of density gradient isolation of the enzyme from insect tissues.

B. Further Purification Procedures

The microsomes obtained can be further purified by extraction with detergent such as Lubrol or sodium dodecyl sulfate. Activation is maximal under specific conditions of protein concentration, detergent concentration, pH, and temperature. No published work is available concerning microsomes from insect tissues. Jørgensen (1974) gave a detailed method for rat kidney microsomes that could be adapted for use with insect material.

The only further purification step on microsomal Na^+,K^+-ATPase from insect tissue is that described by Jenner and Donnellan (1976) using Lubrol W-X extracted microsomes. The pellet obtained was resuspended in water, mixed with an equal volume of glycerol, and centrifuged at 120 000 *g* for 30 min at 4°C. The bulk of the original microsomal ATPase activity was solubilized and showed a 20-fold increase in activity over the original; it was stable during storage at −20°C for 4 weeks.

3. Lyophilization

Peacock (1979, 1981a, 1981b) and Tolman and Steele (1976) have used the convenient technique of lyophilization to make preparations with Na^+,K^+-ATPase activity from insect tissues. The tissue is dissected out and then homogenized in a small volume (1-5 ml) of deionized water using a Potter-Elvehjem homogenizer. The homogenate is rapidly frozen in liquid air and vacuum dried in a freeze drier. The lyophilate can be stored in sealed ampoules at $-20°C$ until required.

Prior to use the lyophilate is transferred to a precooled homogenizer tube containing deionized water (or 30 m*M* histidine buffer pH 7.2 + 1 m*M* EDTA) and reconstituted by homogenization at 0°C. The undispersed material can be sedimented by centrifugation (1000 *g* for 10 min) or by filtration through a Yale No. 30 needle. A final concentration of lyophilate protein of 200–400 μg/ml gives a suitable level of Na^+,K^+-ATPase activity. Peacock (1979) demonstrated that this technique allows the enzyme activity of a single dissected rectum of *Locusta* to be determined. This technique is potentially useful for determining total enzyme activity levels in tissues for comparative studies.

4. Final Remarks on Isolation Procedures

A variety of isolation methods have been attempted with insect tissues. They are all based on well-defined procedures in use with vertebrate tissues and differ mainly in the extent of purification attempted.

The NaI-extracted preparations produce microsomes with low Mg^{2+}-ATPase and high Na^+,K^+-ATPase activities. This makes these microsomes useful for characterization studies. The lyophilization technique is easy and convenient, but a relatively higher Mg^{2+}-ATPase activity is found. However, the two techniques yield Na^+,K^+-ATPase with an equivalent specific activity. The glycerol extraction procedure used by Jenner and Donnellan (1976) for insect "brain" Na^+,K^+-ATPase produced an enzyme with at least twice the specific activity of any derived from the excretory system.

III. Incubation and Assay

1. Assay Conditions

The Na^+,K^+-ATPase is always accompanied by Mg^{2+}-ATPase activity and a differential assay is not possible. Activity is usually determined from the release of Pi in the presence of Mg^{2+}, ATP, Na^+, and K^+ (total ATPase) minus the activity in the same medium containing ouabain (ouabain-insensitive activity). Some workers choose to assay the Mg^{2+}-ATPase activity in the presence of Mg^{2+} and ATP only, and subtract that from the total activity to derive the Na^+,K^+-ATPase.

Both methods of assay may present problems. Several reports suggest Mg^{2+}-ATPase in some tissues is sensitive to monovalent cations, particularly Na^+, at concentrations used to activate the Na^+,K^+-ATPase. Therefore, the Mg^{2+}-ATPase component of the total ATPase activity may not be accurately assayed in a medium containing Mg^{2+} only (Bonting et al. 1964; Izutsu et al. 1974; Schin and Kroeger 1980; Tirri et al. 1979). It is essential to determine whether a significant monovalent cation effect on the Mg^{2+}-ATPase is present because it may well obscure the correct level of Na^+,K^+-ATPase activity. The procedure in which ouabain is included in the complete medium to obtain the ouabain-insensitive (i.e., Mg^{2+}-ATPase) component of the total ATPase avoids any differential monovalent cation effects on the Mg^{2+}-ATPase, but ouabain (1 m*M*) may not completely inhibit the Na^+,K^+-ATPase; thus Na^+,K^+-ATPase activity could be underestimated (Schin and Kroeger 1980; Tolman and Steele 1976). This was found to be so in some insect preparations. Ouabain inhibition is markedly temperature sensitive (Fig. 8.2a); this fact, together with marked species differences in ouabain-sensitivity (Erdmann and Schoner 1973), makes it crucial that appropriate control experiments are carried out. With an uncharacterized preparation it is safest to carry out assays under the following ionic conditions:

1. Mg^{2+}
2. $Mg^{2+} + Na^+ + K^+$
3. $Mg^{2+} + Na^+ + K^+$ + ouabain
4. $Mg^{2+} + Na^+$
5. $Mg^{2+} + K^+$.

In most reports of demonstrations of Na^+,K^+-ATPase from insect sources the authors have not set up conditions that test for either complete ouabain inhibition or a monovalent cation effect on the Mg^{2+}-ATPase. Only in the studies of Anstee and Bell (1975), Rutti et al. (1980), and Schin and Kroeger (1980) have these precautions been taken. In the protocol above the activity in conditions 1, 3, 4 or 5 when subtracted from that in condition 2 should yield comparable activity levels for the Na^+,K^+-ATPase.

The composition of the reaction medium found optimal by most workers is similar to media used with microsomal preparations from vertebrate sources. Some workers have reported that in some preparations atypical [Na^+] and [K^+] are required to activate the enzyme maximally (Table 8.2).

No activity has been found for either the Mg^{2+}-ATPase or Na^+,K^+-ATPase in the absence of Mg^{2+}; as for vertebrate Na^+,K^+-ATPase, a [Mg^{2+}] : [ATP] ratio of 1:1 has been found optimal for most preparations (Rubin et al. 1980). The apparent Michaelis constant (appK_m) for ATP for the Na^+,K^+-ATPase has been reported for several insect species and tissues (Anstee and Bell 1978; Jenner and Donnellan 1976; Peacock et al. 1976; Schin and Kroeger 1980) and lies between 0.1 and 0.5 m*M*. Thus, it is of the same order of magnitude as that described for the enzyme from vertebrate sources (Kline et al. 1971).

The choice of buffer for assay may not be critical. However, Peacock (1981a) and Rivera (1975) both described a higher activity for the Na^+,K^+-ATPase in histidine- or imidazole-based buffers. Sprecht and Robinson (1973) found a similar effect in a preparation from mammalian brain, and they suggested that histidine acts to chelate inhibitory trace heavy metal ions originating either in the medium or preparation.

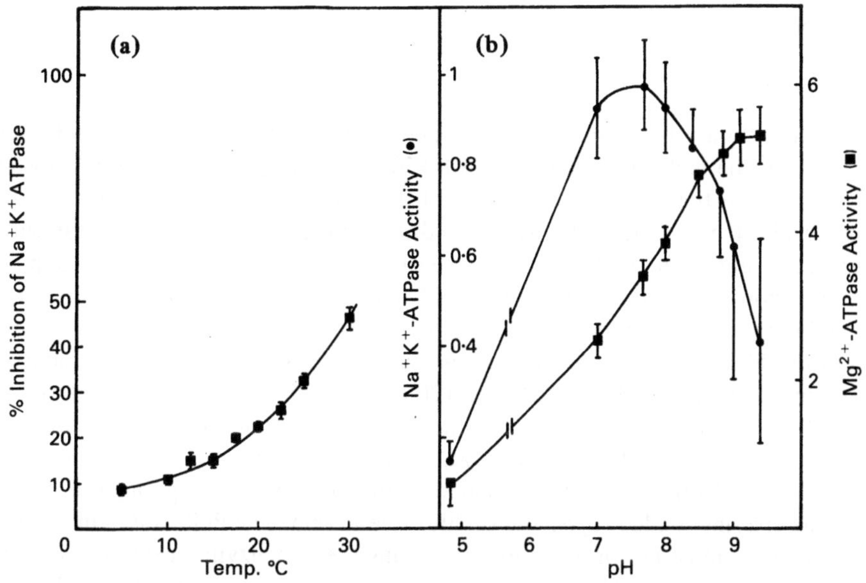

Fig. 8.2. **(a)** Effect of temperature on ouabain inhibition of Na^+,K^+-ATPase from the rectum of *Locusta.* [Adapted from Peacock AJ (1981a) Further studies of the properties of locust rectal Na^+-K^+-ATPase, with particular reference to the ouabain sensitivity of the enzymes. Comp Biochem Physiol C **68**:29–34] **(b)** Effect of pH on Na^+,K^+-ATPase (●) and Mg^{2+}-ATPase (■) activity of the salivary glands from the late fourth instar larvae of *Chironomus thummi* [Adapted from Schin K, Kroeger H (1980) $(Na^+ + K^+)$ATPase activity in the salivary gland of a dipteran insect, *Chironomus thummi.* Insect Biochem **10**: 113–117]

A pH dependency curve is shown in Fig. 8.2b for both the Mg^{2+}-ATPase and Na^+,K^+-ATPase from *Chironomus* salivary gland. As can be seen from Fig. 8.2b and Table 8.1, the pH optimum for the latter enzyme from various insect tissues lies on the alkaline side of neutrality. The Mg^{2+}-ATPase usually has a pH optimum of greater than 8.

Table 8.2. Composition of Reaction Medium (in m*M*) Used in Studying Na^+,K^+-ATPase from Insect Tissues

Mg^{2+}	ATP	Na^+	K^+	EDTA	pH	Reference
2	2	60	5	0.1	7.2	Rutti et al. (1980)
2	4	40	20	–	7.4	Kapoor (1980)
2	2	60	10	–	7.6	Schin and Kroeger (1980)
2–5	20	65	65	0.4	7.6	Rubin et al. (1980)
5	3	60	10	0.5	7.6	Grasso (1967)
5	5	100	22	–	7.6	Cheng and Cutcomp (1975)
4	3	110	20	0.1	7.3	Peacock (1981b)

2. Assay Procedures

A. Enzyme

All media should be preincubated at reaction temperature before the reaction is started by the addition of enzyme (50–150 μg protein). The reaction is allowed to proceed so that no more than 10% of the substrate is hydrolyzed. It is essential to check that under the assay conditions used the reaction is linear with respect to both time and enzyme concentration. The usual method of stopping the reaction is to add a four- to five-fold volume of ice-cold 12% trichloroacetic acid (w/v) to each reaction tube. The resultant mixture is stood in an ice bath until the Pi released is determined. It is usually necessary to remove precipitated protein by centrifugation at 0°–4°C (2000 *g* for 15 min) because it interferes with the Pi assay. All experimental and control tubes must be kept ice-cold because the low temperature slows the acid hydrolysis of the ATP present. For this reason control tubes must be incubated in the same way as experimental tubes but the enzyme is added *after* the trichloroacetic acid. The Pi liberated in experimental tubes must be corrected by subtracting the nonenzymatically liberated Pi in the control tubes. The latter also compensates for any Pi present in the enzyme preparation as a contaminant.

B. Inorganic Phosphate

Two methods are described. The first is the most common and is based on the method of Fiske and Subbarow (1925). The following reagents are used: (1) 5% (w/v) ammonium molybdate; (2) 2.5 *M* H_2SO_4; equal volumes of these two are mixed before use to give acid molybdate; (3) Fiske and Subbarow reducing agent (1-amino-2-naphthol-4-sulphonic acid; Sigma Chemical Co.), freshly prepared as per purchasing instructions in deionized water (other reducing agents such as $FeSO_4$ (40 mg/ml) can be used); and (4) Pi standard solution (20 μg/ml) serially diluted to give a range of standard solutions.

To a 1-ml aliquot of the Pi containing solution, 1 ml acid molybdate is added. It is allowed to stand for *exactly* 10 min before 0.25 ml Fiske and Subbarow reducing agent is added. After *exactly* 20 min the extinction of the blue color produced is read at 660 nm. The timing must be standard for all tubes for only then are they comparable. A curve drawn using data from the standard solutions allows the optical density of the unknowns to be converted into nanomoles Pi.

The second method of Pi assay is preferred by us as it is simpler, requiring one pipetting, and in most instances it is not necessary to remove protein by centrifugation. This method was developed by Atkinson et al. (1973). The reagents used are (1) 1% (w/v) cirrasol ALN-WF in deionized water (ICI Dyestuffs Division, Manchester, U.K.), and (2) 1% (w/v) ammonium molybdate in 0.9 *M* H_2SO_4; equal volumes of these solutions are mixed immediately before use.

Two volumes of the cirrasol acid molybdate are added to the reaction medium. This serves to terminate the reaction and provide the chromogenic agent for phosphate assay. The resulting mixture is allowed to stand at room temperature for 5 min and the extinction of the color produced is read at 390 nm. Reagent control and standard phosphate tubes are also necessary for the reasons given in Sect. III.2.A. The calibration curve is linear up to 0.6 μmol/ml.

Several other methods for Pi assay are commonly used but none are more convenient nor more accurate. Accurate kinetic studies require enzyme activity to be followed by the release of ^{32}P from γ-labeled ^{32}P-ATP. This technique has not been applied to kinetic studies on the enzyme from an insect tissue source; however ^{32}P release has been used by Rubin et al. (1980, 1981) in an enzyme assay.

3. Synthesis of Tris Salt of ATP

It is usual to purchase ATP as the Na salt, and the presence of this ion in the reaction medium is not always desirable. Therefore a method is described for the preparation of the Tris salt of ATP.

Suspend 20 g Dowex 50-X8 resin in 1 liter of 1 *M* HCl and stir for 30 min. Allow the beads to settle and pour off the acid. Wash the beads with deionized water until the washings are at pH 3-4. These washed beads can be stored wet at 4°C until required.

To about 5 g prepared resin, in a Buchner funnel lined with several layers of Whatman No. 1 paper, add 50 ml ATP (sodium salt) solution slowly and let it drain through without suction. Pass the ATP solution through the beads in this way six times. Then wash the resin with 150 ml deionized water and collect with the ATP solution. It is now in the acid form. Titrate with 1 *M* Tris and bring the pH to 7.2. Then add deionized water to a final volume of 250 ml. The resulting Tris-ATP solution can be stored frozen at −20°C until required. The weight of the ATP (sodium salt) taken should be calculated to give the required concentration in the final volume, that is, 250 ml.

4. Use of Inhibitors

The use of inhibitors of Na^+,K^+-ATPase has been a powerful tool in elucidating aspects of the reaction mechanism of the enzyme. In studies on insect Na^+,K^+-ATPases only the cardiac glycoside, ouabain, has been extensively used.

Ouabain (strophanthin-G) is preferred because of its relatively high solubility in aqueous solution (about 1 g in 75 ml at room temperature). Other glycosides require the use of alcohol or organic solvents, which complicates the interpretation of results because the solvents are usually inhibitory too (Rubin et al. 1980).

A. Ouabain Sensitivity Studies

The sensitivity studies are usually carried out by including ouabain in the incubation media and determining residual activity. The range of concentrations chosen is usually 10^{-8}–10^{-3} *M.* The residual activity is plotted as a percentage of the uninhibited rate against the $-\log_{10}$ ouabain (molar) concentration. A typical plot is shown in Fig. 8.3a and the negative logarithm of the molar concentration resulting in 50% inactivation (pI_{50}) can be determined graphically.

An alternative method is to preincubate the enzyme in a medium containing ATP, Mg^{2+}, Na^+, and various ouabain concentrations. Under these conditions ouabain binding is maximal (see Sect. I). After preincubation with ouabain the ATPase reaction is started by the addition of K^+ (where appropriate) in a ouabain solution. The residual activity is determined and plotted as before. It is usually found, when ouabain is present during incubation only, that pI_{50} values are smaller (Akera 1971) and that inhibition is dependent on assay time

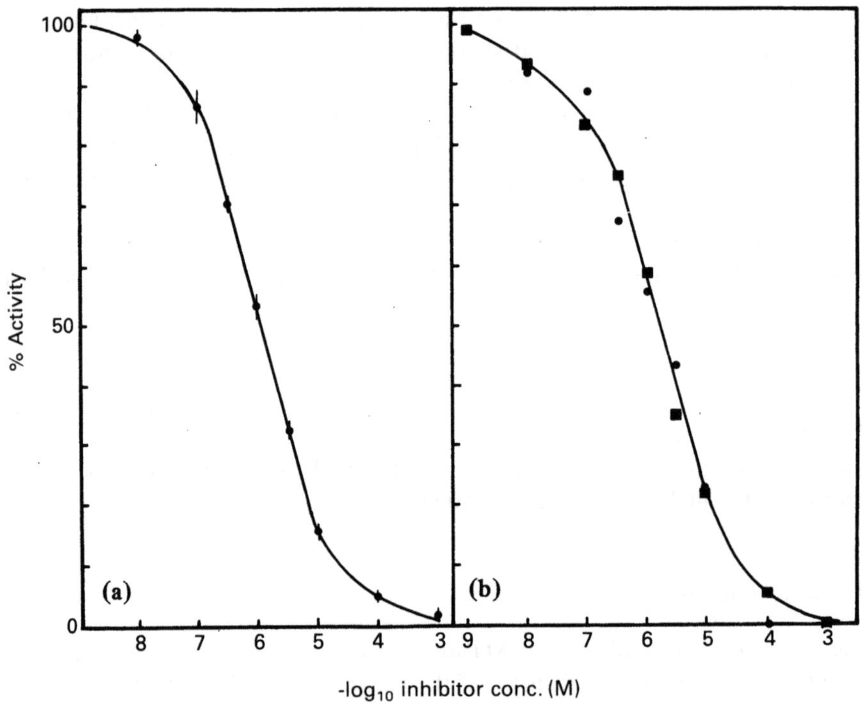

Fig. 8.3. **(a)** Effect of different concentrations of ouabain on Na^+,K^+-ATPase activity in microsomal preparations from Malpighian tubules of *Locusta*. **(b)** Effect of different concentrations of sodium orthovanadate on Na^+,K^+-ATPase in microsomal preparations from Malpighian tubules of *Locusta*. Effect of orthovanadate was determined following preincubation with the enzyme (●) and during incubation alone (■). [Courtesy of Dr. K. Bowler and Dr. J. H. Anstee]

(Schwartz et al. 1975). Rivera (1975) is the only worker to report that an insect Na^+,K^+-ATPase is less sensitive to ouabain in incubation as compared with pre-incubation conditions. Any differences obtained for pI_{50} values between the pre-incubation and incubation methods probably result from K^+ modification of oubain binding (see Sect. I). The K^+ "partially" protect the enzyme against ouabain inhibition (Bonting 1970; Glynn and Karlish 1975). This is also evident in studies on the enzyme from insect tissues (e.g., Jenner and Donnellan 1976; Rutti et al. 1980; Schin and Kroeger 1980). In view of this it is necessary to maintain K^+ concentrations as low as possible, but still consistent with optimal activity, when using ouabain as an inhibitor of Na^+,K^+-ATPase either biochemically or physiologically (Anstee and Bowler 1979).

Temperature is also a modifier of ouabain inhibition, as shown in Fig. 8.2a. In general, a given ouabain concentration produces a lower level of inhibition at a lower temperature (Donkin and Anstee 1980; Peacock 1981a; Peacock et al. 1976). Ahmed and Judah (1965) suggested that this temperature effect results from an enhanced K^+ affinity at low temperatures.

As can be seen from Table 8.1, the calculated pI_{50} for the enzyme from most insect sources lies between 7 and 5. This is very similar to values quoted for mammalian enzymes (Erdmann and Schoner 1973). In most cases the dose-response curve for ouabain is monophasic and may be described in terms of simple uncompetitive inhibition. Rubin et al. (1980) showed, in a preparation from *Manduca sexta,* that the dose responsivity to ouabain is a complex function of ouabain (and strophanthidin) concentration. Similar reports exist for preparations from mammalian tissues (Robbins and Baker 1977; Schwartz et al. 1975), indicating complex inhibitory kinetics or a sensitive and an insensitive population of Na^+,K^+-ATPase sites.

B. Ouabain Binding Studies

Na^+,K^+-ATPase binds ouabain specifically according to the mass law equation:

$$O + R \underset{k_{+1}}{\overset{k_{-1}}{\leftrightharpoons}} OR$$

where O is the ouabain concentration, R is the receptor concentration, OR is the ouabain-receptor complex concentration, and k_{+1} and k_{-1} are the association and dissociation rate constants, respectively.

The formation of ^3H-ouabain–enzyme complex is a time- and temperature-dependent process following second-order kinetics. Thus, if the initial receptor concentration b and the initial ouabain concentration a are known, the association rate constant k_{+1} can be calculated:

$$k_{+1} = \frac{2.303}{(a-b)t} \log \frac{b(a-x)}{a(b-x)}$$

where x is the amount of ouabain bound to the receptor after reaction time t (Erdmann and Hasse 1975; Erdmann and Schoner 1973).

The dissociation of the ouabain–receptor complex is reported to follow first-order kinetics and to be highly temperature dependent (Erdmann and Hasse 1975). At 0°C the ouabain–receptor complex is very stable. The dissociation rate constant (k_{-1}) can be calculated from the exponential decay of ouabain binding.

The Michaelis constant or dissociation constant K_D can be calculated from the ratio of the dissociation rate constant k_{-1} and the association rate constant k_{+1}:

$$K_D = k_{-1}/k_{+1}$$

Alternatively, the dissociation constant can be measured directly from the equilibrium binding of ouabain as a function of ouabain concentration. The latter approach enables the maximal number of binding sites to be determined approximately. The data are normally plotted according to Scatchard (1969); the intercept with the ordinate gives the maximal number of receptor sites, and K_D can be calculated from the slope. In addition, a straight line plot indicates that there is only one type of receptor with high affinity for ouabain (Erdmann and Hasse 1975). Curved lines in Scatchard plots may occasionally be encountered. Erdmann and Schoner (1973) reported the latter in their studies on ouabain binding to ox brain membranes. They suggested that this may indicate a homotrope cooperative effect of ouabain on its receptor in this tissue.

Ouabain binding studies can also be used to determine the turnover number of the Na^+,K^+-ATPase if it is assumed that there is one ouabain molecule bound per pump site. The maximum amount of ouabain bound per milligram protein is determined in conjunction with the assay of Na^+,K^+-ATPase activity. Specific Na^+,K^+-ATPase activity (Pi released per milligram protein per minute) is then related to the amount of ouabain bound per milligram protein and molecular activity (i.e., turnover number) expressed as Pi released per site per minute. Some researchers have attempted to calculate the number of pump sites per cell. Most such estimations have been carried out on vertebrate erythrocytes, where cell numbers can be readily determined (Erdmann and Hasse 1975; Joiner and Lauf 1978). However, Harms and Wright (1980) attempted to make such an estimation for rat intestinal epithelial cells on the basis that each such cell contains 2.5×10^{-7} mg protein. Once the number of pump sites per cell is known and the turnover number of the pump determined, one can calculate the number of Na^+ transported out of each cell, assuming 3 Na^+ transported per cycle of the pump.

To date, few studies report the use of ^{3}H-ouabain in insect tissue preparations. Fristrom and Kelly (1976) and Jungreis and Vaughan (1977) studied ouabain binding to the imaginal discs of *Drosophila melanogaster,* and midgut and nerve in three lepidopteran species, respectively. Unfortunately, both these studies were carried out under conditions that are inappropriate for determining maximal binding due to the inclusion of K^+ in the incubation media. This cation should be excluded in such studies as it is known to inhibit ouabain binding substantially (Whittam and Chipperfield 1975; see also Sect. I). Similarly, the method used by Komnick and Achenbach (1979) to study ouabain binding to whole rectum of *Aeshna cyanea* was suboptimal because Mg^{2+} was absent from the incubation media.

A number of methods have been developed in association with studies on vertebrate tissue preparations that appear to be appropriate for use with insect material (e.g., Charnock et al. 1977; Hansen 1971; Harms and Wright 1980). However, the most suitable composition of the incubation media is likely to vary with the source and specific activity of the preparation. Binding of ouabain is dependent on the concentration of the receptors (expressed as milligrams protein) and on the initial concentration of the drug (Heller and Beck 1978). Similarly, the temperature of incubation is likely to be important because, as has already been pointed out, the formation of the ouabain-enzyme complex is temperature dependent. Thus Harms and Wright (1980) report that ouabain binding to rat intestinal epithelial membranes was too rapid at 37°C for the rate to be accurately determined. No such difficulty was observed by Hansen (1971), who used plasma membrane fragments from ox brain.

Preliminary studies in this laboratory (Bowler and Anstee, unpublished data) indicate that the following method is suitable for determining ouabain binding to microsomal preparations of Malpighian tubules of *Locusta*. The method involves a rapid Millipore filtration procedure similar to that used elsewhere (Hansen 1971; Harms and Wright 1980). First, 1-2 mg microsomal enzyme protein is incubated at 30°C in 10 ml of a medium containing 5 m*M* $MgCl_2$, 2 m*M* EDTA, 100 m*M* NaCl, 3 m*M* ATP in 20 m*M* imidazole-HCl (pH 7.2), and various concentrations of ^{3}H-ouabain (0.1-10 μM). The latter are obtained by dilution of uniformally labeled ^{3}H-ouabain with unlabeled ouabain to yield ^{3}H-ouabain of 0.5-50.0 Ci/mol specific activity. Then 1-ml samples are removed for binding analysis at various times after the addition of the enzyme preparation and rapidly filtered on Sartorius 0.45-μm filters by suction. Sampling should be continued up to 1 h so that equilibrium binding can be obtained.

The filters are then washed with 3 separate 4 ml aliquots of a cold (0°-4°C) washing medium whose composition is identical to that of the incubation medium but without ATP or radioactivity. This should be adequate to effect removal of unbound ^{3}H-ouabain; to ensure that this is so, liquid scintillation counting is done for samples of filtrate. Charnock et al. (1977) recommended prewashing the filters with washing medium to negate nonspecific binding to the filters. Following filtration and washing, the filters are dissolved in scintillation cocktail and assayed by liquid scintillation counting. The amount of bound ouabain can be calculated from the specific radioactivity used in each experiment and is expressed per milligram protein. Nonspecific ouabain binding is determined by running a parallel set of incubations in which 1-10 m*M* unlabeled ouabain is also present in the medium. Specific binding is obtained by subtraction of the ouabain bound nonspecifically.

To determine the release of ^{3}H-ouabain from the ^{3}H-ouabain-receptor complex, the amount of membrane suspension needed is allowed to bind to

^{3}H-ouabain until equilibrium is attained. At this time, an excess of unlabeled ouabain (1 m*M*) is added and the amount of bound ^{3}H-ouabain is followed with time (Godfraind et al. 1980). Alternatively, the method described by Erdmann and Schoner (1973) may be preferred. They allowed the required amount of membrane preparation to bind ^{3}H-ouabain until equilibrium was attained. The membranes were then centrifuged down at 80 000 *g* for 30 min and the pellet resuspended in 0.01 *M* imidazole-HCl (pH 6.5). Dissociation of this ^{3}H-ouabain-receptor complex was then followed with time in incubation media containing unlabeled ouabain (1 m*M* at pH 7.2).

Whilst Mg^{2+} is essential to the ouabain-binding reaction at least two different binding conformations of the Na^+,K^+-ATPase have been distinguished experimentally by observing the effects of Na^+ on ouabain-binding under different ligand conditions (Tobin and Sen 1970). In the presence of a nucleotide, such as ATP or ADP, Na^+ stimulates the rate at which ouabain is bound by the enzyme ("nucleotide" conditions). In contrast, with Mg^{2+} alone or Mg^{2+} in the presence of Pi ("Mg^{2+}, Pi" conditions), Na^+ inhibits the binding rate (Siegel and Josephson 1972). Recently, Rubin et al. (1981) have carried out a quantitative study on the binding of ouabain to the Na^+,K^+-ATPase in homogenates prepared from brain of *Manduca sexta* and bovine brain cortex under "nucleotide" and "Mg^{2+}, Pi" conditions, using techniques similar to those referred to earlier. They found that the number of binding sites measured was not affected by the different ligand conditions and that the rate of dissociation of the receptor-ouabain complex was 10-fold faster in the *Manduca sexta* preparation than in that from bovine brain. This higher K_D for ouabain-binding in the insect preparation was a contributory factor to its lower sensitivity to ouabain inhibition. Furthermore in both preparations and under both binding conditions K^+ decreased the affinity of ouabain to the Na^+,K^+-ATPase.

C. Orthovanadate

In several studies on the Na^+,K^+-ATPase from vertebrate tissues, orthovanadate has been shown to be a potent inhibitor of the enzyme (Cantley et al. 1977); but it is not a specific inhibitor, for other classes of ATPase are also inhibited (O'Neal et al. 1979). The lack of specificity appears to result from the action of vandate as a transition-state analogue of phosphate; consequently, enzymes that hydrolyze phosphate ester bonds are likely targets for inhibition (Cantley et al. 1978). No studies on the effect of vandate on either the activity of Na^+,K^+-ATPases from insect tissues or on transport of ions across insect epithelial tissues have been published.

Preliminary studies in our laboratory (Kalule-Sabiti et al., unpublished data) have determined the pI_{50} for orthovanadate for inhibition of Na^+,K^+-ATPase from Malpighian tubules of *Locusta*. Microsomes were prepared according to the method in Sect. II, and were either preincubated or incubated with solutions containing orthovanadate (10^{-9}–10^{-3} *M*), in exactly the manner described for ouabain inhibition (Sect. III.4.A). A typical result is shown in Fig. 8.3b. As can be seen, this inhibitor has about the same potency as ouabain in the locust preparation, with a pI_{50} of about 6. No difference was found in the sensitivity of the enzyme between incubation and preincubation with orthovanadate. One precaution we find necessary is to use only freshly made solutions of vanadate in inhibition studies.

IV. Methods for Localization of Na^+,K^+-ATPase in Cells

1. Histochemical and Cytochemical Techniques

Considerable problems exist with the preparation of membrane fractions of a sufficiently defined composition that would permit predictions of the localization of the Na^+,K^+-ATPase. Alternative methods have been sought to establish the cellular site(s) of this enzyme. In the earliest attempts with insect preparations, a cytochemical technique based on the Wachstein and Meisel (1957) method was used. However, this method has been subject to considerable criticism because the deposition of product was not ouabain sensitive (Marchesi and Palade 1967; Mizuhira et al. 1970). The major criticism of the Wachstein-Meisel method concerns the presence of lead salts in the incubation medium, used as a trapping agent for Pi released at the enzyme sites. It has been argued that lead salts inhibit the Na^+,K^+-ATPase (Komnick and Achenbach 1979) and also catalyze nonenzymatic hydrolysis of ATP. Furthermore, the fixation procedures recommended are also likely to inhibit the Na^+,K^+-ATPase (Ernst and Philpott 1970). It seems, therefore, that the use of glutaraldehyde fixation and the use of lead as the trapping agent are inappropriate in studies on Na^+,K^+-ATPase localization. Since the Na^+,K^+-ATPase is only one of a family of enzymes capable of hydrolyzing ATP, it is imperative that appropriate control conditions are used so that the specificity of the localization can be unambiguously related to Na^+,K^+-ATPase activity.

2. Method of Ernst

This method relies on the ability of the Na^+,K^+-ATPase to hydrolyze artificial substrates such as *p*-nitrophenyl phosphate (*p*NPP) in a K^+-dependent, ouabain-sensitive reaction (see Sect. I). Inorganic phosphate and *p*-nitrophenol are the products.

The phosphate released is precipitated by strontium at the site of hydrolysis, and an assessment of enzyme activity can be made by spectrophotometric measurement of the *p*-nitrophenol produced (Ernst et al. 1980). In this way Ernst was able to show that enzyme activity was retained after mild fixation in formaldehyde or with mixtures of glutaraldehyde (0.25%) with formaldehyde (1%). The reaction products were dependent on K^+ and Mg^{2+} in the incubation medium and were ouabain sensitive. Control incubations must also be set up to determine the extent of alkaline phosphatase activity, which also hydrolyzes *p*NPP. The usual control is to replace *p*NPP with β-glycerophosphate as the substrate or to include an inhibitor of alkaline phosphatase (cysteine or levamisole; Firth 1980) in the incubation medium containing *p*NPP. Strontium is chosen as the capture ion for Pi released, as it is less inhibitory of the ATPase than is lead. Following the incubation procedure, the strontium phosphate is converted to lead phosphate by treatment with lead nitrate (for electron microscopy), which can be further treated with ammonium sulfide to deposit lead sulfide for use in light microscopy (Ernst 1972a, 1975; Ernst and Mills 1977).

This technique has been successfully applied to localize Na^+,K^+-ATPase in a variety of tissues (Dibona and Mills 1979). However, insect studies have met with only limited success. Peacock (1976b) and Kalule-Sabiti (personal communication) have applied this technique to studying Na^+,K^+-ATPase localization in the rectal pads and Malpighian tubules, respectively, (Fig. 8.4) of *Locusta migratoria.*

The tissue is fixed for 30 min at 4°C in freshly prepared 3% paraformaldehyde in 0.1 *M* sodium cacodylate buffer (pH 7.5) containing 0.25 *M* sucrose. The tissue is then rinsed in 0.1 *M* sodium cacodylate-sucrose (pH 7.5) and incubated for 60 min in medium containing 20 m*M* $SrCl_2$, 10 m*M* KCl, 10 m*M* $MgCl_2$, and 5 m*M* *p*NPP (disodium salt) in 0.1 *M* Tris-HCl (pH 9.0). Four control incubations are carried out: (1) without *p*NPP and (2) in the presence of 1 m*M* ouabain in the incubation medium; (3) without K^+; and (4) with β-glycerophosphate substituting for *p*NPP to assess alkaline phosphatase activity. Following incubation the tissues are rinsed with 0.1 *M* Tris-HCl (pH 9.0) and then treated with 2% $Pb(NO_3)_2$. They are then washed and postfixed for 15 min with 1% osmium tetroxide. Osmicated tissues are dehydrated in alcohol and embedded in a suitable epoxy resin as described by Ernst (1972a, 1975). Both Peacock (1976b) and Kalule-Sabiti (personal communication) found the results to be somewhat inconclusive. In both tissues, a K^+-stimulated component of *p*NPPase activity was apparent and distinct from alkaline phosphatase; however, it was inconsistently inhibited by 1 m*M* ouabain. Similar results were reported by Ernst (1975) for rat kidney cortex.

This technique appears to be unsuitable for use with some tissues; the reasons are not entirely clear why such inconsistent results are obtained. Ernst (1972b) showed that strontium alters the sensitivity of the K^+-*p*NPPase to ouabain. He reported that a 50-fold increase in ouabain concentration was required to cause 50% inhibition of the Na^+,K^+-ATPase in the presence of strontium. A further complication is that strontium at the levels used is also an inhibitor of the enzyme. Komnick and Achenbach (1979) reported that the K^+-*p*NPPase of larval rectum of *Aeshna cyanea* was also inhibited by 20 m*M* strontium used in incubation. A further complication was that fixation in 3% paraformaldehyde also inhibited the K^+-*p*NPPase, and that these inhibitory effects were more pronounced on the K^+-*p*NPPase as compared to the nonspecific ouabain-insensitive *p*NPPase.

It may well be that the technique of Ernst is suitable only where a tissue is rich in Na^+,K^+-ATPase sites. In such a tissue there may well be sufficient K^+-*p*NPPase activity remaining uninhibited to permit histochemical localization, whereas in tissues less rich in Na^+,K^+-ATPase, the uninhibited activity may be too low to be resolved.

Guth and Albers (1974) recommended a modification to the Ernst method that includes dimethyl sulfoxide instead of strontium in the incubation medium. The P_i released is thought to be captured by Mg^{2+}, and subsequently it is converted to a colored precipitate by treatment with cobalt chloride (Ernst et al. 1980). This modification is useful in light microscopy only, and has been successfully applied to kidney tubules where the staining was K^+ dependent and ouabain sensitive (Guth and Albers 1974).

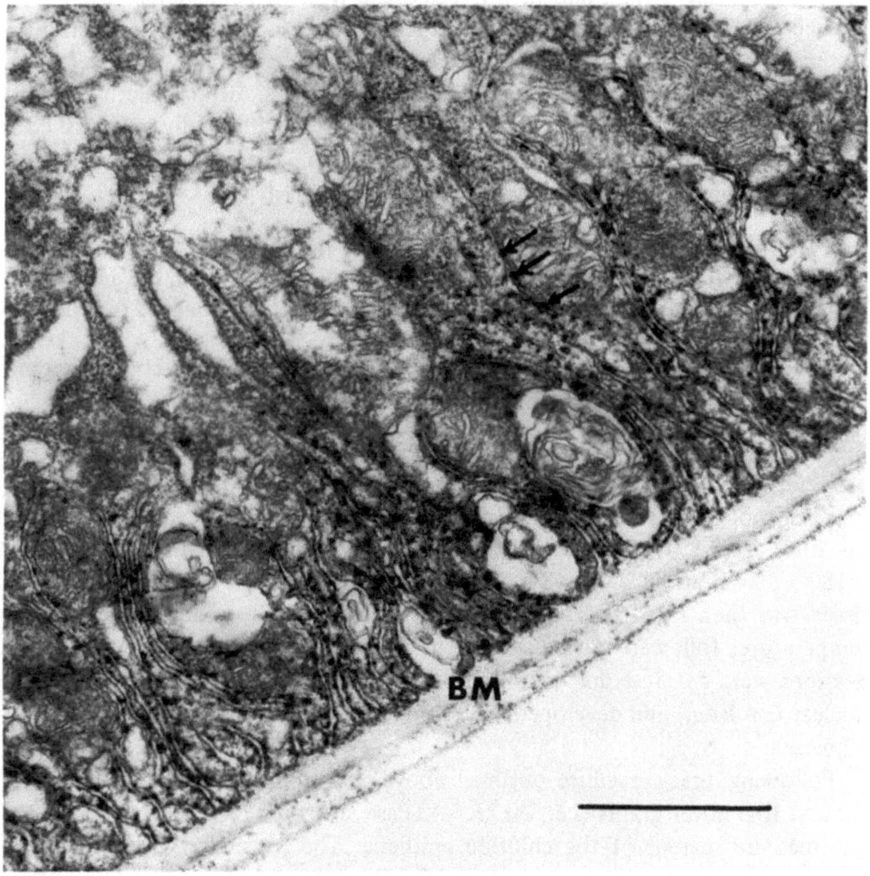

Fig. 8.4. Electron micrograph showing the localization of Na^+,K^+-ATPase sites (arrows) along the infoldings of the basal cell membrane of the Malpighian tubules of *Locusta* as revealed by the *p*NPPase cytochemical procedure of Ernst. BM, basement membrane; scale 1 μm. [Courtesy of Mrs. J. Kalule-Sabiti]

3. Localization Using ^{3}H-Ouabain Binding

Localization of ^{3}H-ouabain binding sites by autoradiography represents perhaps the most promising means of identifying the distribution of Na^+,K^+-ATPase in tissues and cells. The validity of this method depends on the specificity of ouabain binding to the Na^+,K^+-ATPase enzyme. This technique was first introduced by Stirling (1972), who examined the autoradiographic localization of ^{3}H-ouabain binding sites in freeze-dried sections of rabbit intestine. Since this time, numerous researchers have applied this technique to studying vertebrate tissues (see review by Ernst and Mills 1980). In all cases, tissues were exposed to ^{3}H-ouabain followed by rinses in ouabain-free (or "cold" ouabain) Ringer's solutions to remove unbound label. The tissue was then frozen by immersion in a suitable freezing mixture (e.g., liquid propane cooled to −175°C with liquid nitrogen; Stirling 1972), and then freeze-dried. Komnick and Achenbach (1979) applied this method to studying Na^+,K^+-ATPase localization in rectum of *Aeshna cyanea* (Fig. 8.5). Recta were removed from the larvae by dissection in an ice-cold isolation medium containing 125 m*M* NaCl, 50 m*M* sucrose, and 10 m*M* Tris–HCl (pH 7.2). They were then cut open longitudinally and washed in three changes (5 min each) of the above medium. The recta were then transferred to one of the following solutions and kept for 1 h at 4°C:

Solution A: the isolation medium containing 12 μ*M* unlabeled ouabain
Solution B: the isolation medium containing 12 μ*M* ^{3}H-ouabain (25 μCi/ml)
Solution C: the isolation medium containing 12 μ*M* ^{3}H-ouabain (25 μCi/ml) plus 12 m*M* unlabeled ouabain
Solution D: 125 m*M* KCl, 50 m*M* sucrose, 10 m*M* Tris–HCl (pH 7.2) containing 12 μ*M* ^{3}H-ouabain (25 μCi/ml).

Following incubation, solutions A–C were removed by six washes for 10 min each with the isolation medium. Solution D was similarly removed by washing in a solution of similar composition without ouabain. The recta were then blotted on filter paper, transferred to aluminum foil, and freeze-dried for 1 h at −180°C, 24 h at −120°C, 48 h at −80°C, and, finally, 1 h at +30°C. The dried tissue was then fixed overnight at 5 torr in osmium tetroxide vapor at room temperature, followed by embedding in Spurr's (1969) low-viscosity medium. Sections were cut at 1 μm, mounted on glass slides, covered with Kodak NTB_2 nuclear emulsion, and developed with Kodak D-19 after exposure times of up to 12 weeks.

Following the procedure outlined above, Komnick and Achenbach (1979) showed that silver grains (i.e., Na^+,K^+-ATPase sites) are located in the basal and intermediate regions of the chloride epithelia. The possibility that these autoradiographic results are due to nonspecific ^{3}H-ouabain binding can be dismissed on the basis of evidence from control incubations (in solutions C and D). The grain distribution in these controls was not significantly above background and corresponded to that of the control (i.e., in solution A). Mills and Dibona (1980) similarly concluded that all the grains revealed in their autoradiographic studies reflect functional Na^+ pump sites.

The incubation temperature (4°C) used by Komnick and Achenbach (1979) is much lower than has been applied elsewhere (18°C–20°C; Karnaky et al.

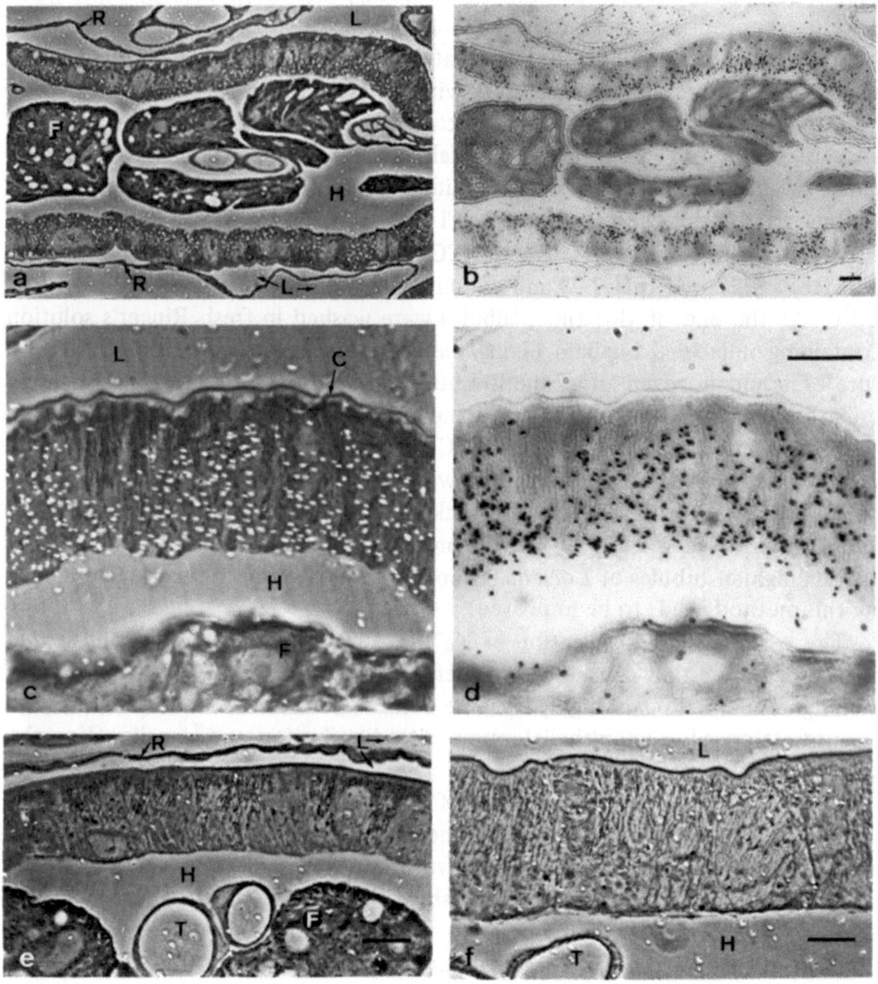

Fig. 8.5. Localization of Na^+,K^+-ATPase sites (silver grains) in the rectum of larval *Aeshna cyanea* by ^{3}H-ouabain autoradiography. Photomicrographs are of sections following incubation in various solutions (see text). **(a)–(d)** Solution B. **(e)** Solution C. **(f)** Solution D. Phase contrast was used except for parts b and d, which were photographed under bright-field illumination. C, cuticle of chloride epithelium; F, fat body; H, hemocoel; L, rectal lumen; R, respiratory epithelium; T, trachea; scale 10 μm. [Reproduced with permission from Dr. H. Komnick and Dr. U. Achenbach]

1976; 37°C, Stirling 1972). Shaver and Stirling (1978) showed that ouabain binding to renal medullary slices of rabbit was negligible at 4°C, while uptake at 25°C was substantial. Furthermore, the uptake at 25°C was almost doubled by increasing the temperature to 35°C. Erdmann and Schoner (1973) reported a similar increase in ouabain binding to microsomes over the same temperature range. This effect of temperature must clearly be taken into account when selecting appropriate incubation temperatures (Fig. 8.2a). Although ouabain-binding was successfully demonstrated at 4°C in larval rectum of *Aeshna cyanea,* higher temperatures may be necessary with different insect species.

Donkin (1981) used a slight modification of this procedure to show the presence of Na^+,K^+-ATPase sites in the Malpighian tubule cells of *Locusta.* Malpighian tubules attached to a collar of midgut-hindgut junction were incubated in a Ringer's solution consisting of NaCl 100 m*M*, KCl 8.6 m*M*, $CaCl_2$ 2 m*M*, $MgCl_2$, 8.5 m*M*, NaH_2PO_4 4 m*M*, $NaHCO_3$ 4 m*M*, NaOH 11 m*M*, glucose 34 m*M*, and HEPES 25 m*M* (pH 7.2) and containing 25 μCi/ml of ^{3}H-ouabain for 1 h at 30°C. At the end of this time, tubules were washed in fresh Ringer's solution containing unlabeled ouabain (1 m*M*) and then frozen rapidly in a 50:50 mixture of liquid nitrogen and 2-methyl butane. Fresh-frozen sections were subsequently cut at 12 μm in a cryostat, transferred to glass slides, fixed in formol saline (10 ml formalin:7 ml 10% NaCl:83 ml distilled water), air dried, and coated with Ilford K5 emulsion. Following a suitable exposure time (6–8 weeks at 4°C), they were developed in Kodak D-19 and stained with toluidine blue. This technique adequately reveals the presence of ouabain binding in the cells of the Malpighian tubules of *Locusta,* although the quality of the sections obtained by this method needs to be improved.

The advantage of localization of Na^+,K^+-ATPase by autoradiography is that it is a technique that can be used under conditions in which the tissue is functioning. The method is highly specific for Na^+,K^+-ATPase and lends itself to quantification of intraepithelial pump site distribution and density (Karnaky et al. 1976; Mills et al. 1977).

The autoradiographic localization of ^{3}H-ouabain requires that the glycoside binds in an essentially irreversible manner with the tissue enzyme and that it is not extracted by subsequent preparative procedures. Therefore, not all tissues may be suitable for this approach: ouabain binds so tightly to avian salt gland enzyme that it cannot be extracted by washing in unbound glycoside, whereas ouabain bound to toad urinary bladder is rapidly extracted by such washing (Ernst and Mills 1980). Shaver and Stirling (1978) showed that, in renal tubules of rabbit, the rate of dissociation of the ouabain-inhibitor complex is temperature dependent. Thus, bound ouabain was not significantly extracted by low-temperature washes (about 0°C), whereas increasing the temperature to 25°C resulted in a significant washout of the labeled glycoside. Accurate localization can only be achieved with tissues that have a relatively high affinity to ouabain.

Ernst and Mills (1980) and Ernst et al. (1980) noted that tissue dehydration, prior to embedding for microtomy, must be accomplished by freeze-drying. This represents a major disadvantage in this technique because disruption of tissue fine structure by small ice crystals means that this method is unsuitable for use at the electron microscopic level. The main argument for freeze-drying seems to be that whereas bound ouabain resists dissociation in aqueous rinses, it is highly soluble in organic solvents such as ethanol (Ernst and Mills 1980; Ernst et al. 1980). However, ouabain is less soluble in alcohol than in water. It would seem, therefore, that the argument in favor of dehydration by freeze-drying should be reexamined. Donkin (personal communication) showed that conventional dehydration by a graded series of alcohol solutions did not prevent autoradiographic localization of ^{3}H-ouabain binding sites in Malpighian tubules of *Locusta* (Fig. 8.6). This finding seems to suggest that electron microscopic autoradiography is

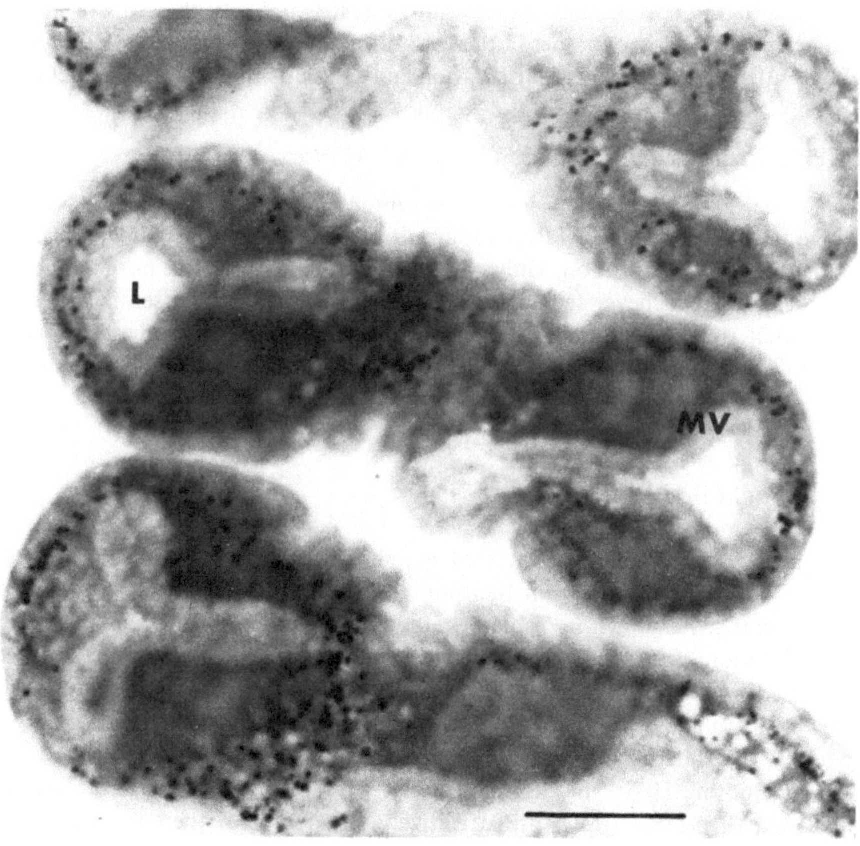

Fig. 8.6. Photomicrograph showing that localization of Na$^+$,K$^+$-ATPase (silver grains) in the Malpighian tubule cells of *Locusta* is possible by ^{3}H-ouabain autoradiography following conventional dehydration and epoxy resin embedding. Section 1–2 μm thick. L, lumen of tubule; MV, microvilli; scale 20 μm. [Courtesy of Dr. J. Donkin]

feasible. However, Stirling (1976) reported that ^{3}H-ouabain was rapidly extracted from ultrathin sections by water in the knife-boat. This would seem to rule out the use of this technique where such sections are necessary. Nevertheless, the loss of labeled glycoside from 1-2 μm sections was less than 20%. It may be, therefore, that newer electron microscopes, which permit 1-μm sections to be studied, will enable this technique to be applied at the fine-structural level.

4. Immunoferritin Localization of Na^+,K^+-ATPase

Kyte (1976) introduced the powerful technique of immunoferritin localization of Na^+,K^+-ATPase sites. It requires specific antibodies to be raised against a purified enzyme. This technique has been successfully applied in vertebrate tissues, where it allows high-resolution localization, by electron microscopy, of ferritin attached to the specific antibodies. However, it is unlikely that this technique will be applied to insect tissues in the near future because it would be difficult to obtain sufficient material for purification and subsequent antibody production.

5. Final Remarks on Localization of Na^+,K^+-ATPase

There are currently three generally accepted methods of localizing Na^+,K^+-ATPase: K^+-dependent *p*NPPase localization, ^{3}H-ouabain autoradiography, and immunoferritin localization. Only the first two techniques have been applied to insect tissues. Whichever technique is eventually chosen, it is essential to establish that it is actually capable, biochemically, of localizing the activity in a given tissue. Thus far, only Komnick and Achenbach (1979) have done this for an insect tissue.

VI. Concluding Remarks

Bonting (1970) summarized the evidence showing a clear relationship between the Na^+,K^+-ATPase and cation transport in tissues from vertebrate and invertebrate animals. As shown in Table 8.1, the enzyme is present not only in membrane fractions from the central nervous system of various insect species, but also in such ion-transporting epithelia as gill, salivary gland, midgut, rectum, and Malpighian tubules. These enzymes, where they have been characterized, show a remarkable similarity in properties to the enzyme from vertebrate brain and kidney (Bonting 1970). In addition, the level of activity obtained is substantial, indicating that these tissues are a relatively rich source of the enzyme. Several attempts at cytochemical and histochemical localization point to the enzyme being present in the basal and lateral plasma membranes in these cells (Figs. 8.4–8.6)—a localization consistent with the proposed involvement of the enzyme in several of the models produced for fluid and ion transport. These various biochemical and cytochemical studies suggest a major role for Na^+,K^+-ATPase in the functioning of fluid- and ion-transporting epithelial cells in insects, as elsewhere (Anstee and Bowler 1979).

References

Abu-Hakima RE, Davey KG (1979) A possible relationship between ouabain-sensitive (Na^+-K^+) dependent ATPase and the effect of juvenile hormone on the follicle cells of *Rhodnius prolixus*. Insect Biochem **9**:195–198

Ahmed K, Judah JD (1965) On the action of strophanthin G. Can J Biochem **43**:877–880

Akera T (1971) Quantitative aspects of the interaction between ouabain and (Na^+-K^+) activated ATPase *in vitro*. Biochim Biophys Acta **249**:53–62

Albers RW (1967) Biochemical aspects of active transport. Annu Rev Biochem **36**:727–756

Albers RW, Koval G, Siegal GJ (1968) Studies on the interaction of ouabain and other cardioactive steroids with sodium-potassium-activated adenosine triphosphatase. Mol Pharmacol **4**:324–336

Allen JC, Lindenmeyer GE, Schwartz A (1970) An allosteric explanation for ouabain-induced time-dependent inhibition of sodium, potassium-adenosine triphosphatase. Arch Biochem Biophys **141**:322–328

Anstee JH, Bell DM (1975) Relationship of Na^+-K^+-activated ATPase to fluid production by Malpighian tubules of *Locusta migratoria*. J Insect Physiol **21**:177–184

Anstee JH, Bell DM (1978) Properties of Na^+-K^+-activated ATPase from excretory system of *Locusta*. Insect Biochem **8**:3–9

Anstee JH, Bowler K (1979) Ouabain-sensitivity of insect epithelial tissues. A review. Comp Biochem Physiol A **62**:763–769

Atkinson A, Gatenby AD, Lowe AG (1973) The determination of inorganic orthophosphate in biological systems. Biochim Biophys Acta **320**:195–204.

Bonting SL (1970) Sodium-potassium activated adenosine triphosphatase and cation transport. In: Bittar EE (ed) Membranes and ion transport, Vol 1, pp 257–363. Wiley, New York

Bonting SL, Hawkins NM, Canady MR (1964) Studies of sodium-potassium activated adenosine triphosphatase. Biochem Pharmacol **13**:13–22

Caldwell PC, Hodgkin AL, Keynes RD, Shaw TI (1960) Partial inhibition of the active transport of cations in the giant axons of *Loligo*. J Physiol (Lond) **152**:591–600

Cantley LC, Josephson L, Warner R, Yamagisawa M, Lechene C, Guidotti G (1977) Vanadate is a potent (Na-K) ATPase inhibitor in ATP derived from muscle. J Biol Chem **252**:7421–7423

Cantley LC, Cantley LG, Josephson L (1978) A characterization of vanadate interactions with the (Na, K) ATPase. Mechanistic and regulatory implications. J Biol Chem **253**:7361–7368

Charnock JS, Simonson LP, Almeida AF (1977) Variation in sensitivity of cardiac glycoside receptor characteristics of (Na^+-K^+) ATPase to lipolysis and temperature. Biochim Biophys Acta **465**:77–92

Cheng EY, Cutcomp LK (1975) The ATPase system in American cockroach muscle and nerve cord. Insect Biochem **5**:421–427

Dibona DR, Mills JW (1979) Distribution of Na^+-pump sites in transporting epithelia. Fed Proc **38**:134–143

Donkin JE (1981) Some effects of insect hormones on Na^+, K^+-ATPase and fluid secretion by the Malpighian tubules of *Locusta migratoria* L. Doctoral dissertation, University of Durham, Durham UK

Donkin JE, Anstee JH (1980) The effect of temperature on the ouabain-sensitivity of Na^+–K^+-activated ATPase and fluid secretion by the Malpighian tubules of *Locusta*. Experientia **36**:986–987

Erdmann E, Hasse W (1975) Quantitative aspects of ouabain binding to human erythrocyte and cardiac membranes. J Physiol (Lond) **251**:671–682.

Erdmann E, Schoner W (1973) Ouabain–receptor interactions in (Na^+-K^+) ATPase preparations from different tissues and species. Biochim Biophys Acta **307**:386–398

Ernst SA (1972a) Transport adenosine triphosphatase cytochemistry. II. Cytochemical localization of ouabain-sensitive, potassium-dependent phosphatase activity in the secretory epithelium of the avian salt gland. J Histochem Cytochem **20**:23–38

Ernst SA (1972b) Transport adenosine triphosphatase cytochemistry. I. Biochemical characterization of a cytochemical medium for the ultrastructural localization of ouabain-sensitive, potassium-dependent phosphatase activity in the avian salt gland. J Histochem Cytochem **20**:13–22

Ernst SA (1975) Transport ATPase cytochemistry: Ultrastructural localization of potassium-dependent and potassium-independent phosphatase activities in rat kidney cortex. J Cell Biol **66**:586–608

Ernst SA, Mills JW (1977) Basolateral plasma membrane localization of ouabain-sensitive sodium transport sites in the secretory epithelium of the avian salt gland. J Cell Biol **75**:74–94

Ernst SA, Mills JW (1980) Autoradiographic localization of tritiated ouabain-sensitive sodium pump sites in ion transporting epithelia. J. Histochem Cytochem **28**:72–77

Ernst SA, Philpott CW (1970) Preservation of Na-K-activated and Mg-activated adenosine triphosphatase activities of avian salt gland and teleost gill with formaldehyde as fixative. J Histochem Cytochem **18**:251–263

Ernst SA, Riddle CV, Karnaky KJ (1980) Relationship between localization of Na^+-K^+-ATPase, cellular fine structure, and reabsorptive and secretory electrolyte transport. In: Bronner F, Kleinzeller A (eds) Current topics in membranes and transport, Vol 13, pp 355–385. Academic Press, New York

Firth JA (1980) Reliability and specificity of membrane adenosine triphosphatase localizations. J Histochem Cytochem **28**:67–71

Fiske CH, Subbarow Y (1925) The colorimeter determination of phosphorus. J Biol Chem **66**:375–400

Fristrom JW, Kelly L (1976) Effects of β-ecdysone and juvenile hormone on the Na^+/K^+-dependent ATPase in imaginal disks of *Drosophila melanogaster*. J Insect Physiol **22**:1697–1707

Garay RP, Garrahan PJ (1973) The interaction of sodium and potassium with the sodium pump of red cells. J Physiol (Lond) **231**:297–325

Garrahan PJ, Rega AF (1972) Potassium activated phosphatase from human red blood cells. The effects of *p*-nitrophenylphosphate on cation fluxes. J Physiol (Lond) **223**:595–617

Gilbert JC, Wyllie MG (1975) Effects of prostaglandins on the ATPases of synaptosomes. Biochem Pharmacol **24**:551–556

Glynn IM (1968) Membrane adenosine triphosphatase and cation transport. Br Med Bull **24**:165–169

Glynn IM, Karlish SJD (1975) The sodium pump. Annu. Rev. Physiol **37**:13–55

Godfraind T, De Pover A, Dutete DT (1980) Identification with potassium and vanadate of two classes of specific ouabain binding sites in a (Na^+-K^+)ATPase preparation from the guinea-pig heart. Biochem Pharmacol **29**:1195–1199

Grasso A (1967) A sodium and potassium stimulated adenosine triphosphatase in the cockroach nerve cord. Life Sci **6**:1911–1918

Guth L, Albers RW (1974) Histochemical demonstration of (Na^+-K^+)-activated adenosine triphosphatase. J Histochem Cytochem **22**:320–326

Hansen O (1971) The relationship between g-strophanthin-binding capacity and ATPase activity in plasma membrane fragments from ox brain. Biochim Biophys Acta **233**:122–132

Harms V, Wright EM (1980) Some characteristics of Na/K-ATPase from rat intestinal basal lateral membranes. J Membr Biol **53**:119–128

Heller M, Beck S (1978) Interactions of cardiac glycosides with cells and membranes—Properties and structural aspects of two receptor sites for ouabain in erythrocytes. Biochim Biophys Acta **514**:332–347

Hodgkin AL, Keynes RD (1955) Active transport of cations in giant axons from *Sepia* and *Loligo*. J Physiol (Lond) **128**:28–60

Ilenchuk TT, Davey KG (1982) Some properties of Na^+-K^+ ATPase in the follicle cells of *Rhodnius prolixus*. Insect Biochem **12**:675–679

Izutsu KT, Siegel IA, Brisson DL (1974) The effect of ionic strength on a Mg^{2+} ATPase and its relevance to the determination of (Na^+-K^+) ATPase. Biochim Biophys Acta **373**:361–368

Jenner DW, Donnellan JF (1976) Properties of the housefly head sodium and potassium-dependent adenosine triphosphatase. Insect Biochem **6**:561–566

Joiner CH, Lauf PK (1978) Ouabain binding and potassium transport in young and old populations of human red cells. Membr Biochem **1**:187–202

Jørgensen P (1974) Isolation of the (Na^+-K^+) ATPase. In: Fleisher S, Packer L (eds) Methods in enzymology, Vol 32. Academic Press, New York

Jørgensen P, Skou JC (1971) Purification and characterisation of the (Na^+-K^+) ATPase. Biochim Biophys Acta **233**:366–380

Jungreis AM, Vaughan GL (1977) Insensitivity of lepidopteran tissues to ouabain: Absence of ouabain-binding and Na^+-K^+ ATPases in larval and adult midgut. J Insect Physiol **23**:503–509

Kapoor NN (1980) Relationship between gill Na^+ K^+-activated ATPase activity and osmotic stress in the plecopteran nymph, *Paragnetina media*. J Exp Zool **213**:213–218

Karnaky K, Kinter LB, Kinter WB, Stirling CE (1976) Teleost chloride cell II. Autoradiographic localization of gill Na, K-ATPase in killifish *Fundulus heteroclitus* adapted to low and high salinity environments. J Cell Biol **70**:157–177

Kline MH, Hexum TD, Dahl JL, Hokin LE (1971) Studies on the characterisation of the sodium-potassium transport adenosine triphosphatase. Arch Biochem Biophys **142**:781–787

Komnick H, Achenbach U (1979) Comparative biochemical, histochemical and autoradiographic studies of Na^+/K^+-ATPase in the rectum of dragonfly larvae (Odonata, Aeshnidae). Eur J Cell Biol **20**:92–100

Kyte J (1976) Immunoferritin determination of the distribution of (Na^+-K^+) ATPase over the plasma membranes of renal convoluted tubules I. Distal segment. J Cell Biol **68**:287–303

Lindenmayer GE, Schwartz A (1973) Nature of the transport adenosine triphosphatase digitalis complex. IV. Evidence that sodium-potassium competition modulates the rate of ouabain interaction with (Na^+-K^+)-adenosine triphosphatase during enzyme catalysis. J Biol Chem **248**:1291–1300

Marchesi VT, Palade GE (1967) The localization of Mg-Na-K-activated adenosine triphosphatase on red cell ghost membranes. J Cell Biol **35**:385–404

Matusi H, Schwartz A (1966) Purification and properties of a highly active ouabain-sensitive Na^+,K^+-dependent adenosine triphosphatase from cardiac tissue. Biochim Biophys Acta **128**:380–390

Mills JW, Dibona DR (1980) Relevance of the distribution of Na^+ pump sites to models of fluid transport across epithelia. In: Bronner F, Keinzeller A (eds) Current topics in membranes and transport, Vol 13, pp 387–400. Academic Press, New York

Mills JW, Ernst SA, Dibona DR (1977) Localization of Na^+-pump sites in frog skin. J Cell Biol **73**:88–110

Mizuhira V, Amakawa T, Yamashina S, Shirai N, Utida S (1970) Electron microscopic studies on the localization of sodium ions and sodium-potassium-activated adenosine triphosphatase in chloride cells of eel gills. Exp Cell Res **59**:346–348

Nakao T, Tashima Y, Nagano K, Nakao M (1965) Highly specific sodium-potassium-activated adenosine triphosphatase from various tissues of rabbit. Biochem Biophys Res Commun **19**:755–758

Norris DM, Cary LR (1982) Properties and subcellular distribution of Na^+,K^+-ATPase and Mg^{2+}-ATPase in the antennae of *Periplaneta americana*. Insect Biochem **11**:743–750

O'Neal SG, Rhoads SB, Racker E (1979) Vanadate inhibition of sarcoplasmic reticulum Ca^{2+} ATPase and other ATPases. Biochem Biophys Res Commun **89**:845–850

Peacock AJ (1976a) The effect of corpus cardiacum extracts on the ATPase activity of locust rectum. Insect Biochem **22**:1631–1634

Peacock AJ (1976b) Distribution of Na^+-K^+-activated ATPase in the alimentary tract of *Locusta migratoria*. Insect Biochem **6**:529–533

Peacock AJ (1977) Distribution of Na^+-K^+-activated ATPase in the hindgut of two insects *Schistocerca* and *Blaberus*. Insect Biochem **7**:393–395

Peacock AJ (1978) Age dependent changes in Na^+-K^+, activated ATPase activity of locust rectum. Experientia **34**:1546–1547

Peacock AJ (1979) A comparison of two methods for the preparation of Mg^{2+}-dependent, (Na^++K^+) stimulated ATPase from the locust rectum. Insect Biochem **9**:481–484

Peacock AJ (1981a) Further studies of the properties of locust rectal Na^+-K^+-

ATPase, with particular reference to the ouabain sensitivity of the enzyme. Comp Biochem Physiol C **68**:29–34

Peacock AJ (1981b) Distribution of (Na^++K^+)-ATPase activity in the mid- and hindguts of adult *Glossina morsitans* and *Sarcophaga nodosa* and the hindgut of *Bombyx mori* larvae. Comp Biochem Physiol A **69**:133–136.

Peacock AJ (1982) Effects of sodium transport inhibitors on diuresis and midgut (Na^++K^+)-ATPase in the tsetse fly *Glossina morsitans*. J Insect Physiol **28**:553–558

Peacock AJ, Bowler K, Anstee JH (1972) Demonstration of a Na^++K^+-Mg^{2+}-dependent ATPase in a preparation from hindgut and Malpighian tubules of two species of insect. Experientia **28**:901–902

Peacock AJ, Bowler K, Anstee JH (1976) Properties of Na^+-K^+-dependent ATPase from the Malpighian tubules and hindgut of *Homorocoryphus nitidulus vicinus*. Insect Biochem **6**:281–288

Piccione W, Baust IG (1977) Effects of low temperature acclimation on neural Na^+-K^+ dependent ATPase in *Periplaneta americana*. Insect Biochem **7**:185–189

Post RL, Sen AK, Rosenthal AS (1965) A phosphorylated intermediate in adenosine triphosphate-dependent sodium and potassium transport across kidney tubules. J Biol Chem **240**:1437–1445

Proverbio F, Condrescu-Guidi M, Whittembury G (1975) Ouabain-insensitive Na^+ stimulation of an Mg^{2+}-dependent ATPase in kidney tissue. Biochim Biophys Acta **394**:281–292

Rivera ME (1975) The ATPase system in the compound eye of the blowfly, *Calliphora enythrocephala* (Meig.). Comp Biochem Physiol **52**:227–234

Robbins AR, Baker RM (1977) (Na,K)ATPase activity in membrane preparations of ouabain resistant HeLa cells. Biochemistry **16**:5163–5168

Robinson JD, Flashner MS (1979) The (Na^+-K^+)-activated ATPase. Enzymatic and transport properties. Biochim Biophys Acta **549**:145–176

Rubin AL, Clark AF, Stahl WL (1980) Sodium, potassium stimulated adenosine triphosphatase in the nerve cord of the hawk moth, *Manduca sexta*. Comp Biochem Physiol B **67**:271–275

Rubin AL, Clark AF, Stahl WL (1981) The insect brain (Na^++K^+)-ATPase. Binding of ouabain in the hawk moth, *Manduca sexta*. Biochim Biophys Acta **649**:202–210

Rutti B, Schlunegger B, Kaufman W, Aeschlimann A (1980) Properties of the Na, K-ATPase from the salivary glands of the ixodid tick *Amblyomma hebraeum*. Can J Zool **58**:1052–1059

Scatchard G (1969) The attractions of proteins for small molecules and ions. Ann NY Acad Sci **51**:660–672

Schatzmann HJ (1953) Herzglykoside als slemmstoffe für den activen kalium and natrium transport durch die erythrocytenmembran. Helv Physiol Pharmacol Acta **11**:346–354

Schin K, Kroeger H (1980) (Na^++K^+)ATPase activity in the salivary gland of a dipteran insect, *Chironomus thummi*. Insect Biochem **10**:113–117

Schwartz A, Lindenmayer GE, Allen JC (1975) The sodium potassium adenosine triphosphatase: Pharmacological, physiological and biochemical aspects. Pharmacol Rev **27**:3–134

Sen AK, Tobin T, Post RL (1969) A cycle for ouabain inhibition of sodium- and potassium-dependent adenosine triphosphatase. J Biol Chem **244**:6596–6604

Shaver JLF, Stirling C (1978) Ouabain binding to renal tubules of the rabbit. J Cell Biol **76**:278–292

Siegel GJ, Josephson L (1972) Ouabain reaction with microsomal (Sodium-*plus*-Potassium)-activated adenosinetriphosphatase. Europ J Biochem **25**: 323–335

Skou JC (1957) The influence of some cations on an ATPase from peripheral nerves. Biochim Biophys Acta **23**:394–401

Skou JC (1965) Enzymatic basis for active transport of Na^+ and K^+ across cell membrane. Physiol Rev **45**:596–617

Skou JC (1975) The (Na^++K^+) activated enzyme system and its relationship to transport of sodium and potassium. Q Rev Biophys **7**:401–434

Skou JC, Butler KW, Hansen O (1971) The effect of magnesium, ATP, Pi and sodium on the inhibition of the (Na^++K^+)-activated enzyme system by g-strophanthin. Biochim Biophys Acta **241**:443–461

Sprecht SE, Robinson JD (1973) Stimulation of the Na^+-K^+-dependent adenosine triphosphatase by amino acids and phosphatidylserine, chelation of trace metal inhibitors. Arch Biochem Biophys **154**:314–323

Spurr AR (1969) A low viscosity epoxy resin embedding medium for electron microscopy. J Ultrastruct Res **26**:31–43

Stirling CE (1972) Radioautographic localization of sodium pump sites in rabbit intestine. J Cell Biol **53**:704–714

Stirling CE (1976) High resolution autoradiography of ^{3}H-ouabain binding in salt transporting epithelia. J Microsc (Oxf) **106**:145–157

Tirri R, Tumola P, Bowler K (1979) The presence of Na^+ ATPase activity associated with mammalian brain microsomal preparations. Int J Biochem **11**: 43–48

Tobin T, Sen AK (1970) Stability and ligand sensitivity of [^{3}H]ouabain binding to (Na^++K^+)-ATPase. Biochim Biophys Acta **198**:120–131

Tolman JH, Steele JE (1976) A ouabain-sensitive, (Na^++K^+)-activated ATPase in the rectal epithelium of the American cockroach, *Periplaneta americana*. Insect Biochem **6**:513–517

Vaughan GL, Jungreis AM (1977) Insensitivity of lepidopteran tissues to ouabain: Physiological mechanisms for protection from cardiac glycosides. J Insect Physiol **23**:585–589

Wachstein M, Meisel E (1957) Histochemistry of hepatic phosphatases at a physiologic pH. With special reference to the demonstration of bile canaliculi. Am J Clin Pathol **27**:13–23

Whittam R, Chipperfield AR (1975) The reaction mechanism of the sodium pump. Biochim Biophys Acta **415**:149–171

Yap HH, Cutcomp LK (1970) Activity and rhythm of ATPases in larvae of the mosquito *Aedes aegypti*. Life Sci **9**:1419–1425

Index

Absolute pressure transducers
  as a supplementary device to an oxygen electrode 106
Absorption
  measurement of across everted rectal sacs 32
Absorption mechanisms
  physiology of 89
Accuracy 174
Acetate
  transport of in locust gut 31
Active Ca transport in locust rectum
  preferred substrate for 33
Active Cl transport
  temporary inhibition of 33
Active epithelial transport
  anoxia of 156
Active transport 42, 156
  effect of putative hormones on 42
  inhibitors of 42, 156
Activity coefficient 46
  of Ca, 21
Activity of water (a_w) 69, 73
Adenosine 3′, 5′-monophosphate (cAMP) 25
Adenosine triphosphate *see* ATP
Aedes aegypti 166, 190
  properties of Na^+, K^+-ATPase from 190–191
Aerobic animals
  total energy metabolism in 163
Aeshna cyanea 208, 210
  ATPase in rectum of 210–211
Agar bridge electrodes 48
Agar bridges 15, 36
  sealing of 36
Agla micrometer syringe 10
Albumin
  use as a volume marker 26
Alkaline phosphatase
  inhibitors of 208
Amaranth dye 22
Amaranth red 11
Amblyomma hebraeum (a tick) 192
  ATPase from, 192–193
Amblyomma varigatum (a tick) 81, 84
Amino acids 31, 102
  transport of in locust gut 31
Ammonium
  measurement of 22
Ammonium molybdate 200
AMTF (artificial Malpighian tubule fluid) 30
Anesthesia *see also* narcosis
Anesthesia by chloroform 113
Anesthesia by CO_2 22, 26, 83, 113
Anesthesia by ether 26
Anodic side reactions

Anodic side reactions (*cont.*)
effect of 179
Anomalous freezing
role of glycoproteins in 91
Antifreezing agents
anomalous freezing caused by 91
Antimony electrode 22
Apical cell border
rate of Na and Cl uptake across 33
Apical membrane potential (V_a) 51
Apiezon grease
for leak-free connections 116
Aquafluor 40
Araldite
sealing with 36
Arenivaga investigata 81, 84, 88–90
Artifact caused by preequilibration of rectal sac 34
Artifacts
interference due to HCO_3 causing 47
Artifacts due to vigorous shaking 110
Artificial hemolymph 20
Artificial Malpighian tubule fluid (AMTF)
locust recta perfused with 30
Aseptic work in insect tissue culture 113
Association rate constant 204
Atomic absorption spectrophotometry 21
ATP (adenosine triphosphate)
biochemical energy in the form of 132
effect of ouabain on 146
ATPase
partial purification of from *Manduca sexta* midgut 157
Automatic gamma counter for ^{22}Na 33
Automatic volumetric burettes 105
Autoradiography 211–213
Avian salt gland slices
NADH fluorescence in 153
A_w (activity of water) 69, 73

Baralyme 167–168, 176
absorption of CO_2 by 166, 176
Beehive 163
Beeswax 86
Beeswax-resin mixture 26
Beetle
chronocoulometric respirometry for 175
Beryllium–bronze foil
use in microrespirograph 119–120, 122
Bessel function 53
Bimodal daily rhythm in oxygen consumption 183
Blaberus craniifer 193
ATPase from 192–193
Blaberus discoidalis 184
Black epoxy plug 29
Bombyx Mori 111, 192
ATPase from 192–193
Brain
fluorescence from mitochondrial NADH in 145
Bumblebees
chronocoulometric respirometry for 175

^{14}C
scintillation counting of 40
Cable analysis 52–55
Ca bound to amino acids in locust hemolymph 25
Ca-glutamate complexes
stability constants for 25
Calcium activity coefficient 21
Calliphora erythrocephala 7, 13, 190
properties of Na^+, K^+-ATPase from 190–191
Calomel electrodes 13–16, 30
Camera lucida 112
cAMP (adenosine 3′, 5′ monophosphate) 25, 35, 40, 53
Cannula 10
Capacitance compensation 17
Capacitive coulometry 165–166
Carausius morosus 7
hemolymph of 21
Cardiac glycosides 188, 189
binding of 189
Cartesian diver manometry
disadvantage of 103
Catalytic activity
requirement of Mg^{2+} plus ATP for of ATPase 188
Cathode ray oscilloscope 16
Cathodic current density 165
Centrifugation

isolation of microsomal pellets by 195
procedures in 196
Chamber for mounting recta as flat sheets 44–45
Chironomus
Malpighian tubule of 50
salivary gland of 50
Chironomus thummi 192, 199
ATPase from 192–193
salivary gland from, 199
Chloride transport-stimulating hormone (CTSH) 30
Chloroform narcosis
prohibition of 113
Chronocoulometric measurement
accuracy of 179
Chronocoulometric respirometry 175–179
Chronocoulometry 165, 184
oxygen consumption recorded by 184
^{14}C-inulin
use as a volume marker 26
Circuit diagrams for coulometric measurement 181–182
Circuit diagram for short-circuit of locust rectum 38–39
Cirrasol ALN-WF 200
^{36}Cl 33, 40
test solution containing 33
scintillation counting of 40
Cl electrode 45
CO_2 22, 114
absorption of 114
diffusion of 114
microdiffusion method for total 22
CO_2 anesthesia 22, 26, 83, 113
CO_2 narcosis
prohibition of 113
CO_2 tension
measurement of 22
Coefficient of variation 175
Colophony 86
Common volumetric principle 105
Comparative oxygen consumption data 107
Compensating chamber 105
Compensating thermobarometer 165
Compensating vessel 181
Compensating voltage clamp 36–40
Conductometric method 21
Constant respirometric values 102
Constant-volume differential manometer 106
Copper 165
Copper deposition
systematic discrepancy in 178–179
Copper wire
constant oxygen uptake of 175
Coulometric coefficient (*k*) 174–175, 178–179
Coulometric constant (*k*) 165
Coulometric measurement
effect of mechanical vibration on 169
theory of 164
Coulometric measurement of oxygen consumption
calibration device 173
calibration of 172–175
Coulometric measuring process 164
Coulometric method
accuracy of 174, 179
Coulometric microrespirometer 166–175
Coulometry
examples of application of 183
Counter circuits 182
^{14}C-polyethylene glycol
use as a volume marker 26
Critical equilibrium activity (CEA) 83
Critical point freeze drying 89
CTSH 42
Current injection 51
Current intensity 165
Current-passing electrodes 43
$CuSO_4$ 166–173, 176–178, 180
$CuSO_4$ meniscus
behavior of in microrespirometer 170
Cuticle temperature depression
relationship between evaporation rate and 84
Cyanoacrylic adhesive
use in fixing tensometric sensors to transducer in microrespirograph 120
Cyclamate
Cl replacement by 35
Cyclic AMP (cAMP) 15
Cysteine 208
Cytochrome a, a_3

Cytochrome *a*, a_3 (*cont.*)
amount present in solubilized preparations calculated using extinction coefficient 151
high affinity for oxygen of 134
measurement of 156
redox level of USA measurement of tissue anoxia 152
spectrophotometry of 150
Cytochrome *a*, a_3 in respiratory chain 132, 134
Cytochrome b_5
unknown function of 150
Cytochrome enzymes 101
Cytochrome redox level
anoxia of 156
Cytochrome redox state
quantification of 138

Danaus Plexippus 190
properties of Na^+, K^+-ATPase from 190–191
Datel model DPP-Q7 printer 181
Debye constant 46
Debye–Hückel equation 46
Deoxydiolate
purpose of exposing membranes to 196
Dermacentor andersoni (tick) 166
Dermatophagoides pteronyssinus (European house dust mite) 96
Dermatophagoides farinae (house dust mite) 83, 96
tritium-exchange kinetics of 84
Detecting platinum microelectrode 177
Determining pH in nanoliter volumes 21
Dew point hygrometer 77
Dew point microvoltmeter
measurement of vapor absorption by 94–95
Diaferometry 106
advantages 106
disadvantages 106
Diethylether narcosis
prohibition of 113
Differential manometer
central part of respirometer involves 116
Differential pressure transducer
use in microrespirograph 118–119, 124
Differential respirometers 103
Digital coulometric respirometers 164
Dioxane-based fluors
use of 40
Direct volumetric respirometry
calibration of 104
Dissecting the locust rectum 35–36
Dissociation rate constant 204
Down-counter 178, 181
Double-barreled ion-exchanger microelectrodes 43–46
Double-barreled ion-selective microelectrode 48
Dowex 50-X8 201
Dragonfly rectum 36
Drop volume
calculation of 93
plot of versus reciprocal of vapor pressure lowering 94
Drosophila 163
Drosophila melanogaster 166, 183, 190, 204
imaginal discs of 204
properties of Na^+,K^+-ATPase from 190–191
seven-day oxygen consumption of 183
Dye-coupling 50
Dye-injecting microelectrode 55
Dye injection 50

Edge damage in locust recta 42
EDTA (ethylenediaminetetraacetate) 193, 195
Electrically coupled cells 50–51
Electricity in coulombs
quantity of 165
Electrochemical gradient of an ion
formula for 49
Electrolytic cell
constant current flow through 178
Electrolytic respirometers 106
construction of 106
Electrometer 13, 15–16, 30, 47
Electrometric titration with $AgNO_3$
determination of Cl by 27
Electron microprobe x-ray analysis 91

Electron proble microanalysis 21
Endocrine glands 102
Endogenous substrate 102
Endothermic energy 101
Endothermic oxidations 101
Enzymes 174, 196
  density gradient isolation of 196
  purification of 174
Epithelia 144, 149–150
  cytochrome content of 150
  use of spectrophotometry in 149
  uses of fluorometry in 144
Epithelia in chamber
  ability to properly oxygenate 153
Epithelial permeability 42
  hormones on 42
  inhibitors of 42
Epithelial transport
  techniques used in
    applied to vapor-transporting tissues 96
Epithelial transport and metabolism
  simultaneous alterations in 148
Epithelium
  electrical and chemical gradients across
    control of 41
Epoxy resin 29, 120, 167
  hermetically sealed by 167
  use in fixing tensometric sensors to transducers in microrespirograph 120
Error in weighing caused by hygroscopic absorption 77
Errors
  random (precision) 174
  systematic (accuracy) 174
Errors due to nascent oxygen 181
Errors in humidity control due to temperature fluctuation 71–72
Errors in use of ion-sensitive microelectrodes 54–55
Etching
  platinum wire electrode 166
Etching of metal
  use of $FeCl_3$ solution for 119
Ether anesthetized 26
Ethylacetate narcosis
  prohibition of 113
Ethylene dichloride
  glue made from Plexiglas shavings and 36
Evaporative cooling 73
Evaporative water loss
  studies of 70
Everted rectal sac preparations
  advantages of 33–34
  limitations of 33–34
Exogenous substrate 101–102
Exchange kinetics 71
Expiratory CO_2
  absorption of 114
Explanted fat body
  respiration of 106
Extracellular compartment 187
Extramitochondrial cytochromes
  no evidence for 150
Eyepiece graticule 5

Faraday 165
Faraday's law 163–165, 180–181
  deviations from 181
Fatty acid 102
Fecal equilibrium technique 87–88
Fecal pellets
  water content of 89
$FeSO_4$
  reducing agent 200
Fiber optics 43
Filling solutions for microelectrodes 46
Fiske and Subbarow reducing agent (1-amino-2-naphthol-4-sulphonic acid) 200
Flame photometry 21, 27
  measurement of Na and K by 27
Flat-sheet cable analysis 52–53
Flat sheet preparation 34
Flavoprotein in respiratory chain 132
Flip-flop 178, 181
Flow rate
  choice of in humidity control 76
Fluorescent dye
  injection of 50
Fluorescence signals
  background fluorescence, excitation angle and amount of tissue

Fluorescence signals (*cont.*)
determines magnitude of 136
Formulae for converting osmotic pressures to solvent activity 69
Freezing point depression 69, 92
conversion of to vapor pressure lowering 92
Frog gastric mucosa
cytochrome redox states in 154
energy production and consumption in 143
reduced-minus-oxidized spectrum of 150

Gasometric detection
sensitivity of 170
Gastric mucosa
potassium-sensitive ATPase in 154
Glossina 7
Glossina morsitans 192
ATPase from 192–193
Gluconate 23, 40
replacement of external Cl by 40
use of to replace Cl during flux experiments 23
Glucose–agar medium 163
Glue made from ethylene dichloride and Plexiglas shavings 36
α-Glycerophosphate 102
Glycogen 102, 109
Grace's medium
explanted organisms in 111
Grace's tissue culture medium 108
Grasshopper embryos 109
Gravimetric techniques 70, 74, 94
Gut ligation 22

Half-maximal activation of ATPase 188
HCO_3
insensitivity to of Cl-sensitive microelectrodes 47
Heat-coagulated hemolymph 110
Heat-exchanger apparatus 77–79
Helium glow photometry 21
Hematological slides 112
Hemocoel 19
Hemolymph 20, 110
collection of 20
self-oxidation ability of 110
storage of 20
Hemolymph Ca, Mg bound to amino acids in locust 25
Hemolymph K activity
diurnal variations in 25
Hemolymph phosphate 21
Hexoses 102
High-affinity Na^+ sites
presence in ATPase of 188
High-energy phosphate bonds 101
High impedance amplifier 15–16
Hindgut 19, 189
role in renal function 19
Histidine-based buffer 198
^{3}H-mannitol 33
test solution containing 33
Homorocoryphus nitidulus 192
ATPase from 192–193
Honeybees 163, 175
chronocoulometric respirometry for 175
^{3}H organics
scintillation counting of 40
Hornet
chronocoulometric respirometry for 175
Horsefly maggot 19
^{3}H-Ouabain 205
^{3}H-Ouabain autoradiography 211
^{3}H-Ouabain binding
localization of, 210–214
^{3}H-Ouabain-enzyme complex
formation of 203, 205
^{3}H-Ouabain-receptor complex 205–206
rate of dissociation of 206
House dust mites 83, 96
(*Dermophagoides* species) 96
H_2SO_4 accumulation
effect on calibration of 174
Humidified air
regulated supply of 77
Humidity chambers 73–74
convective circulation in 73
Petri dishes as 74
Humidity-controlling solutions 73
Humidity probes 74
Humidity-regulating insert with a Mettler

ME22 electronic microbalance 75
Humidity sensors 74
Hyalophora cecropia 51, 150, 190–191
properties of Na^+, K^+-ATPase from 190–191
spectrophotometry of midget of 150
Hydrogen peroxide
production of 180
Hydrogen production
reduction to a minimum of 180
Hydrophilic cuticular hairs
high surface tensions of fluid between 89
5-hydroxytryptamine 6, 15–16
Hygroscopic absorption
interference of weighing by 77
Hyperpolarizing current 50

Ideal gas law 164, 172
Ileum 22
Image-intensified vidicon 140
Imaginal discs 204
Imidazole-based buffer 198
Immobilizing insects
method for 113
Impalement artifacts 54
problems with microelectrodes caused by 54
Immunoferratin localization of Na^+, K^+-ATPase 214
Incipient condensation
activities for 73
In Cl-free saline 23
Incubation medium 109
Infrared gas analysis 105
Injection fluid 26
Inorganic phosphate (Pi)
assay for 200–201
Input impedance 47
Input resistance 51
Insect eggs
measurement of respiration kinetics by microrespirograph in 117
Insect epithelia 35, 113
KCl transport across 35
measured volumes of from 10 to 100 nl 113
Insect hemolymph 25
Instantaneous oxygen consumption 166
Intestinal epithelium 102
Intestinal epithelium of adult female *Pyrrhocoris apterus*
scanning microrespirograph showing oxygen consumption of 126
Intermediary metabolism 101
Interference 22
Interference due to HCO_3 47
Intra and extrcellular spaces 107
Intracellular potential differences 16
Intraepithelial pump site
quantification of 212
Inulin 11
In vitro tracer methods 31
Ion-exchanger resins 25
use in short-circuit current experiments 25
Ion-selective microelectrode 42–43
calibration of 43–46
fabrication of 43–46
Ion-sensitive electrodes 25
Ion-sensitive microelectrodes
errors in use of 54–55
Ion transport
dependence of prolonged water absorption by *in vitro* locust rectum on 32
I_{sc} *see* short circuit current 35–42
Isopentane
rapid freezing of tissues by 90
Isoosmotic fluid transport 51

Jamaicana flava 192
ATPase from 192–193
Jet-stream abrasive 50

K^+ active transport of 187
K^+ electrode 45
K^+ intracellular 187
k (coulometric coefficient or constant) 165, 174
Kimax weighing bottle 166
Krebs citric acid cycle intermediates 102

Lanolin
as a lubricant for leak-free connections 116
Laplace's law 170
Latent Na^+, K^+-ATPase
unmasking activity of 196
Leucophaea
hemolymph of 25
Levamisole 208
Ligated locust recta 26
Ligature
human hair as 26
Ligatured abdomen
interference of rectal uptake of water vapor by 86
Limitations
of everted rectal sac preparations 33–34
Liquid index 105
Liquid ion-exchanger microelectrodes 25
Liquid ion-exchangers
used for ion-sensitive microelectrodes 45
Liquid paraffin 8, 15
Liquid scintillation counter 33
Liquid scintillation spectrometry 11
Lithium chloride 50
Lithium hydroxide 22
Liver plasma membrane 189
Localization of Na^+, K^+-ATPase
summary of methods for 214
Locusta
Malpighian tubules of 194
Locusta migratoria 208
properties of Na^+, K^+-ATPase from 190–191
rectal ATPase activity from 199
Locust hemolymph amino acids 25
Locust recta
chamber for mounting as flat sheets 44–45
correction for edge damage in 42
Locust recta perfusion of 28
Locust rectum 23, 36
Low-pass filter 47
LSD (least significant digit) 182
Lubricant for leak-free connections 116
Lubrol
extraction of enzyme with 196
Lubrol W-X (detergent) 196
Lucifer Yellow CH 50, 55
Luer-lock Teflon terminals
use of in microrespirograph 118
Lumen-facing mucosal side 26

Manduca sexta 5, 7, 37, 51, 142, 157, 190, 203
ATP, cytochrome *a*, a_3 and oxidation in midgut of 157
midgut studies of 37
mounting midgut of in spectrophotometry chamber 142–143
properties of Na^+, K^+-ATPase from 190–191
Manometer 166–167, 169, 174
Manometric index
use of isooctane, isodecane, preboiled kerosene, or light mineral oil, etc., stained with Sudan Black as 116
Malpighian tubule 19, 102, 189, 202
dissection of 6–8
localization of ATPase in 208
microsomal Na^+, K^+-ATPase from 194
Na^+, K^+-ATPase from 194, 202–203
of *Schistocerca gregaria* 8
perfusion of 8–9, 12
rapid flushing of 10
role of in excretion and detoxification 6
secretion rates of 6
simultaneous perfusion of 11
size of 5
solute movements across 10
transepithelial potentials of 13
use in epithelial studies 5
Malpighian tubule fluid 20–22, 26
anion deficit in 23
Manometric methods 102
Maximum relative error 175
calculation of 176
Measurement of CO_2 tension 22
Measurement of dissolved oxygen 105
Mechanical vibration
detection improved in presence of 169
Meniscus 169–172

behavior of at $CuSO_4$–air interface 170
Metabolic experiments
duration of 164
Metabolic substrate
heptanoate used as 148
relative effectiveness of 148
Metabolic water 101
Metabolically inert material 107
Methanesulfonate
use of to replace Cl during flux experiments 23
Methyl salicylate 50
clearing solution using 50
Methyl sulfate
use of to replace Cl during flux experiments 23
Mg^{2+}-ATPase 187
Mg-ATP complex 188
Mg-glutamate complexes
stability constants for 25
Mg bound to amino acids in locust hemolymph 25
Michaelis constant (K_m) 192, 198
calculation of 204
Microanalysis methods 21
Microbalance humidity chambers
calibration graph of 76
Microcautery 17
Microdiffusion method for total CO_2 22
Microelectrode
filling solution for 46
holder for 10
penetration of cell by 16
techniques for 42–49
Micromanipulator 9–10
Micrometric burettes 26, 105
Micropipettes 20
Microrespirometer 103, 172
actual precision of 175
function of 168
universal scanning 117
Microsomal Na^+, K^+-ATPase
rate of inhibition of 189
Microvolumetric method 110–113
calculate weight of very small tissue or individual cells by 112
principles of 112
Midgut of tobacco hornworm
changes in redox state of 145
oxidation of NADH in 145
Millipore filtration 205
Mineral oil 20
Miniature differential manometer 105
Miniature inert valve 1XP 118
Mitochondria
list of respiratory states 133
respiratory states of 132–134
techniques for spectrophotometry of 138–140
Mitochondrial cytochromes
double-beam spectrophotometry of 138
Mitochondrial NAD
redox state of 144
Mitochondrial NADH
fluorometry used to monitor redox level of 134
Mitochondrial state transitions
cytochrome redox state used to monitor 134
Mitochondrial uncoupler 155
Monochrometer
use of in rapid-scanning spectrophotometer 139
Moth
chronocoulometric respirometry for 175
Movement artifacts 143
errors in optical measurements caused by 143
errors in oxygen consumption caused by 143
MSD (most significant digit) 182
Mucosal side 20, 33
of gut epithelia 33
Multiple respirometers 105
Multiplexor 139–140
Musca domestica 184, 190
properties of Na^+, K^+-ATPase from 190–191
Muscle 102

^{22}Na
automatic gamma counter for 33
Na^+
activation curves for with membrane

Na^+ (*cont.*)
preparations of Na^+, K^+-ATPase 188
active transport of 187
activity coefficient of 25
intracellular concentration of 187
Na^+ electrode 45
Na^+, K^+-ATPase
accompanied by other ATPase 189
artificial substrates for 188
assay of, 197–201
binding of ouabain to 206
characteristics of 187–188
degree of purification of 194
extraction of 192–197
inhibition of 155
latent activity of 194
limited use of crude homogenates of 194
localization of 207–209, 214
membranes possessing 193
orthovanadate inhibition of 206
ouabain inhibition of 199
preparation of tissue with 192–197
presence of 190–193
properties of 190–191
purification of 194
turnover number of 204
underestimation of activity of 198
Na pool in the rectal fluid of locust 28
NAD (nicotine adenine dinucleotide)
entry into respiratory chain through 132
measurement of redox state of 134–135
non-fluorescence of 134
steady-state redox level of 134
NADH (reduced NAD) 134
fluorescence emission band of 134
fluorescence enhancement of 134
fluorescence signal from 136
microfluorometry of in isolated perfused proximal renal tubules 135
NADH fluorescence
effect of heptanoate addition on 149
effect of ouabain on 146
limitations in interpretation of 136
measurement of 156
problem in quantification of 136
Naphthalimide dye 50
Narcosis *see also* anesthesia
Narcosis by chloroform 113
Narcosis of insects
inhibitory effects on mitochondrial enzymes of 113
Nascent oxygen
dissolution of 181
problems caused by 181
Necturus gallbladder cell 54
Neoprene O-rings 77
Nernst equation 46
Nerve ganglia 102
Net fluid transport
dependence of on ions, metabolic substrate and inhibitors 33
Net transepithelial flux 49
Neurosecretory peptide in locusts
rectal Cl absorption controlled by 30
Nicolsky equation 46
Nicotine adenine dinucleotide (NAD) 132
N-methyl-D-glucamine 25
Nominal saturated vapor pressure 72
Noncompensating voltage clamp 36
Nonlinear uptake kinetics 85
Nonsynchronous temperature fluctuations of $\pm 0.001°C$ 175
Nonvapor absorbers 70
Normal developmental cycles in explanted tissues 111
Nystatin (a polyene antibiotic)
use of 148

O_2 delivery
spectrophotometric study of 144, 152
Ouabain 146, 187, 189, 191, 197, 201–203
binding of 189
localization of in tissues with high affinity for 212
sensitivity of K^+-*p*NPP to 208–209
sensitivity studies of 202
Ouabain (strophanthin-G) 201
Ouabain binding studies 203
Optical comparator 173
Optical measurement
chamber designs for 141
noninvasive nature of 141

Organismic properties
nutritional status of 163
reproductive status of 163
Orthovanadate 202
Ovaries
growing of 111
Oxidative phosphorylation and oxygen consumption
degree of coupling between 155
Oxidative phosphorylation in respiratory chain 132
Oxygen
circuit diagram for measurement of 171–172
electrolytic production of 164–165
measurement of 105
partial pressure of 163
production of unit volumes of 181
solubility of 102
volume of (V_{O_2}) 178
Oxygen consumption 163
bimodal daily rhythm in 183
coulometric measurement of 170–177
daily metabolic cycle in 184
decrease in due to depletion of endogenous substrate 110
decrease in due to ouabain 153
effect of heptanoate addition on 149
effect of ouabain on 146
erroneous reading of 114
errors due to dissolution and diffusion of oxygen in 174
expressing the rate of 111
measurement of by volumetric microrespirometer 114–117
potassium caused rapid acceleration of 145
rate of ($\dot{V}_{O_2}$) 165
recording of in house fly 184
systematic errors in measurement of 180
Oxygen consumption and optical properties
simultaneous monitoring of 144
Oxygen consumption and oxidative phosphorylation
degree of coupling between 155
Oxygen consumption in explanted tissues 108
Oxygen consumption rates
calculation of 107
usually fall within 100-1000 ul $h^{-1}g^{-1}$ live weight 107
Oxygen consumption rates of explanted tissues 107
Oxygen depletion 164
Oxygen diffusion
shortening the path of 110
Oxygen electrode 106, 108
advantage of 106
disadvantage of 106
Oxygen in respiratory vessel
limit of consumption of to not more than 5–10% of the initial amount 114
Oxygen measurement
precision of 179
Oxygen-permeable polyethylene 106
Oxygen pulse generator 177
Oxygen uptake
error in due to copper deposition 165
Ozone
production of 180
trace amounts of 181

Paracellular and transcellular routes 10
Paracellular pathway 11
Paraffin oil 5–6, 9, 13
Paraffin wax 86
Paraformaldehyde
fixation by 50
Paragnetina media 192
ATPase from 192–193
Partial anoxia
explanation for mixed results of measurement of cytochrome redox levels 156
Partial ionic conductance 49
Partial pressure of gases 171
Passive osmotic flow into animals begins when ambient activity exceeds that of the blood 72
PE 10 cannula 30
PE 90 tubing 26
Peltier effect 94
use in a dew point microvoltmeter 94

Perfusion bath 16
Perfusion cannulae 29
Perfusion chamber 15
Pericardial cells 102
Periplaneta americana 189–190
properties of Na^+, K^+-ATPase from 190–191
Perirectal fluid
large amounts of protein in 91
volume-concentration relationship of 92
pH electrode 21
Phenoloxidase
activiation of 110
inhibitor of 110
Phenylthiourea
addition of to reduce self-oxidation 110
Phosphatase
presence of K^+-dependent ouabain-inhibitable 188
Phosphate
transport of in locust gut 31
Phosphate in rectal tissue 21
Phosphorylated intermediate 188–189
Photomultiplier tube 141
Physiological salines 102
composition of 24
problems concerning 25
pI_{50}
for orthovanadate inhibition of Na^+, K^+-ATPase 206
Picapnotherm 22
Pieris brassicae 7
Planimeter
integration of recordings of instantaneous short-circuit current by use of 41
Planimetrically 112
Plasticine 30–31
Platinum anode 165, 166, 169
p-Nitrophenyl phosphate (*p*NPP) 207
as a substrate of ATPase 188
*p*NPP (*p*-Nitrophenyl phosphate)
*p*NPPase
cytochemical procedure of Ernst for 209
PO_2 152
redox level independent of 152
Polargraphic methods 105
Polyethylene tubing 11, 13, 26
Polyvinyl chloride tubing 10
Potassium-selective electrode
measurement of extracellular potassium with 144
Potassium transport
inhibition of 155
Potter–Elvehjem homogenizer 196–197
Precision of measurements
definition of 174
Preequilibrate tracers
alteration of rectal fluid cause by the need to 34
Pressure-detecting circuit
wiring diagram of 177
Primary urine 19
Proctodeal valve 22
Proline
abundance of in rectal tissue 26
Psychrometric measurements 74
Psychrometric method 80
Purification of enzymes 196
Putative hormones
effects of on active transport 42
Pyridine nucleotide (NAD)
entry into respiratory chain through 132
Pyrrhocoris apterus
oxygen consumption in intestinal epithelium of 126

q
electrodynamic expression of 165
electrostatic expression of 165

Rabbit kidney
aerobic respiratory rate in isolated cortical tubules, from 147
cortical tubule suspension from 146
Rabbit proximal renal tubules
reduced-minus-oxidized spectra of 150
Rat liver mitochondrial
reduced minus oxidized spectrum of 137
Rates of tissue respiration 111
Reaction medium for ATPase 199
Rectal absorption in the desert locust
measurement of 27

Rectal Cl absorption 20
Rectal fluid of locust
  Na pool in 28
Rectal lumen
  reabsorption from 19
Rectal pads
  localization of ATPase in 208
Rectal sac
  dissection and mounting of 36
Rectal sac preparations 33
  advantages of 33–34
Rectal tissue proline is the most abundant substrate in 26
Rectum 19
Red blood cell
  ATPase enzymes in 188
Relative deviation in coulometry 174
Relative temperature fluctuation 176
Respiration 101
  effects of added substrated on 109
  effect of inhibitors on 109
  rates of 101
Respiration of isolated fragments of the insect body 109
Respiratory chain 131–140
  flow of electrons through 132
  hemoproteins in 131
  redox state of 141, 155
Respiratory chamber 163
  construction of 176
  designed for insects 166
Respiratory enzymes
  use of spectrophotometry to monitor the redox state of 137
Respiratory states of mitochondria
  definition of 132–134
Respiratory quotient 80, 109
  change in due to switch from carbohydrate to fat oxidation 80
Respiratory rate expressed per unit of weight shows the highest values for the small inactive organ 111
Respiratory rates of insects expressed in units of live weights 112
Respirograph
  calibration of 122–123
  temperature compensation of 124–125
Respirometers 164
Respirometric data 101
  misinterpretation of when expressed per unit protein 111
Respirometric medium
  use of insect hemolymph as special precautions 110
Respirometric methods
  miniaturization of 103
Respirometric techniques 103
Respirometry 102
  capacitive coulometric 166–175
  dironocoulometric 175–179
  microrespirometric 166–179
Restoring active Cl transport 26
Reversible exposure
  technique of 86
Rhodnius prolixus 5, 8, 192
  ATPase from 192–193
  mesothoracic ganglion mass of 6
Rhodotorula minuta (yeast) 180
Ringer's solution 108–113

Salines 20
  activity coefficient of Ca in 21
  amino acid-free 26
  glucose in 25
Salivary glands 102
Salt bridges 13
Sarcophaga bullata 20
Sarcophaga nodosa 192
  ATPase from 192–193
Saturation vapor pressure
  almost directly proportional to temperature fluctuation 72
  definition of 72
  empirical equation relating temperature and 72
Scatchard plots 204
Schistocerca gregaria 19, 26, 192
  ATPase from 192–193
  composition of body fluids collected from 22–23
  measurement of rectal absorption in 27
Secretagogues 154
Sedimentation 197
Semiconductor tensometer type AP-120-6-12

Semiconductor tensometer (*cont.*)
use of in microrespirograph 120
Sequencer 182
Serosal side 20
of gut epithelia 33
Shaking of respiratory chamber
alternative to 110
Shaking racks 105
Short-circuit current 34
changes in 157
monitoring of 155
Short-circuit locust recta
apparatus for 38–39
Side reactions
effect of on cathode 180
Signal-to-noise ratio
multiple scans for improvement of 140
Silastic tubing 43
Silicone grease 15
use of to seal transducer for microrespirograph 120
Silicone oil
insulating electrodes with 43–44
Silicone rubber tubing
compensatory device using 116
Silk ligature 22, 26, 31
locust gut tied with 31
Silkworm
integument of 37
Silkworm midgut
potassium transport in 141
reduced-minus-oxidized spectrum of 150
Ussing chambers to measure tracer fluxes across 34
Skeletal muscle
changes in fluorescence originating from mitochondrial NADH 145
Small tissue
simple volumetric respirometer for 114–115
Small tissue samples
respirometry of 114
Space constant 53
Standard deviations
in coulometry 174
STPD 174
$^{35}SO_4$
scintillation counting of 40
Sodium backflux
estimated rate of 28
Sodium dodecyl sulfate (detergent)
extraction of enzyme with 196
Sodium orthovanadate 202
Solubility of oxygen 102
Spectrophotometry
advantages of 21
Split-beam spectrophotometer
studies on sheet epithelia 141
Spontaneous oxygen consumption
due to activation of phenoloxidase enzymes 110
Squamous epithelial cells 5
Strip chart recorder 30
Substrate dependence of homogenates prepared from different tissues 109
Sudan black 29
Sulfate concentration in slug hemolymph 21
Sulfate in saline 21
Surface area 112
Surface fluorescence
minimization of specular reflection as crucial for measurement of 136
Sylgard 50, 184
Sylgard washers
chamber mounting with 142
Synthetic dyes 102
Systematic errors 180

Tackiwax 22
Teflon membranes 106
Teflon-insulated platinum wire
as a cathode 167
Temperature and saturation vapor pressure
ampirical equation relating 72
Temperature control 173
accuracy of 72–73
Temperature equilibration 171
oxygen consumption preceding 171
Tenebrio molitor 79, 88, 92
osmotic pressures in rectal complex of 75–76
water-exchange compartments of 85

Tensometric detectors
CA-100, CA-110, KWS 73, or M-1000 used as, 121
Tensometric sensor, 118, 119, 121, 126
use of in microrespirograph 118–126
Thermal conductivity detectors 107
Thermal detector 105
Thermistor compensation 106
Thermobarometer (V_T) 165–168, 174–176, 179–180
Thermobia domestica 74, 80–81, 85, 87
Thermocouples 84
Three-way stopcock 26
Tight junctions 51
Timer circuit 182
Tissue culture media 111
Tissue dispersion
common method for 194
Tissue respiration 110
experimental conditions of 109
Titratable acid
measurement of 22
Toad urinary bladder
ouabain bound to 212
Tobacco hornworm
midgut of 157
Total body water
gravimetric determination 83
Total solvent volume
calculation of in either of two vapor pressures 86
Tracheal system 109
Transcellular route 11
Transducer
use of in microrespirograph 118–119
Transepithelial potential 15–16, 29, 34
Transmeniscus pressure 170
Transmission fluorescence
minimization of specular reflection as crucial for measurement of 136
Transport of salts and water across hindgut 26
Transrectal potential 32
Tris salt of ATP
synthesis of 201
Tristate decade counters (DM8552) 181
Tritiated water exchange
measurement of 83
Triton X-100
solubilizing the suspension with 150
Tryptamine 6
Tubule fluid 22
Turnover number 204
Two-channel voltage clamp 37
Tygon tubes
hermetically sealing 176

Ubiquinone in respiratory chain 132
Unidirectional flux 40
Unidirectional flux rate ($J_{1\to 2}$)
formula for 40–41
Unidirectional tracer fluxes 34
Uptake kinetics
study of 71
Ussing chambers 34, 156
Ussing double chamber
flattened version of 142–143

Vacuum aspirator 15
Vapor-absorbing mechanisms
solute-coupled 90
solute-independent 90
Vapor-absorbing systems
use of gravimetric method in 93–94
use of dew point microvoltmeter in 94
dependence on introduction of small amounts of water between 5 and 200 μl 95
Vapor absorption 71
electron microprobe analysis of ultrathin frozen sections for study of 91
inhibition of by occulsion 86
most species must be in a state of water deficit before they begin to absorb vapor 71
restoration of body water reserves by 80
Vapor absorption from the atmosphere 71
Vapor pressure lowering 69, 92, 94
conversion from freezing point depression 92
Vapor uptake
energy dependence of 83

Vaseline 15
Vectoral enzyme 188
V_M 165
V_{O_2} 165
Voltage comparator 171
Voltage divider ratios 49
Voltage follower 171–172
Voltage-sensing microelectrodes 51
Volume of a tissue or organ
  calculation of 112
Volume marker
  ^{14}C-insulin as 26
  ^{14}C-polyethylene glycol as 26
Volumetric microrespirometer 105
  oxygen consumption of small tissues measured by 114–117
  working procedures for 117
Volumetric respirometer 105, 108
Volumetric unit
  of microrespirograph 118–119
V_t *see* transepithelial potential 35, 42
V_T (thermobarometer) 167

Wachstein–Meisel method 207
Warburg method 108–109
  advantages of 103
  disadvantages of 103
  modification of 103
Warburg methods for respirometry
  limitations of 113
Warburg techniques 109–110
Wasp
  chronocoulometric respirometry for, 175
Water
  partial pressure of 164
Water absorption
  difficulty in separating active and passive absorption 83
Water concentrations
  stability of 71
  stability of water concentrations in air depends directly on the precision of temperature control 71
Water-exchange compartments 85
Water potential 69
Water-saturated oil 20
Water tracer studies with tritium
  principal difficulty of 83
Water transporting organs 96
  Malpighian tubules as 96
  midgut as 96
  rectum as 96
  salivary glands as 96
Water uptake rate
  correction for surface losses in calculation of 84
Water vapor
  coefficient of diffusion of 89
Water vapor absorption
  apparatus for 78
  balance techniques for measurement of 80
  study of 70
Water vapor concentration expressed as vapor pressure 69
Water vapor uptake
  investigations of 71
Weight changes
  analysis of non-steady-state 85
Wescor water potential cell
  measurement of water absorption by use of 95
Wescor water potential probe 80

Xenopsylla 84
Xeric environments 71
Xylene/surfactant-based fluors
  use of 40
Xylose 11

MIX
Papier aus verantwortungsvollen Quellen
Paper from responsible sources
FSC® C105338

If you have any concerns about our products,
you can contact us on
ProductSafety@springernature.com

In case Publisher is established outside the EU,
the EU authorized representative is:
Springer Nature Customer Service Center GmbH
Europaplatz 3, 69115 Heidelberg, Germany

Printed by Libri Plureos GmbH
in Hamburg, Germany